Making Environmental Policy

Making Environmental Policy

Daniel J. Fiorino

UNIVERSITY OF CALIFORNIA PRESS
Berkeley · Los Angeles · London

This book was written by the author in his private capacity. No official support or endorsement by the U.S. Environmental Protection Agency is intended or should be inferred. The views expressed in this book are those of the author and not necessarily those of the Environmental Protection Agency or the U.S. government.

University of California Press
Berkeley and Los Angeles, California

University of California Press
London, England

Library of Congress Cataloging-in-Publication Data

Fiorino, Daniel J.
Making environmental policy / Daniel J. Fiorino.
p. cm.
Includes bibliographical references and index.
ISBN 0-520-08918-9 (pbk.)
1. Environmental policy—United States. 2. United States. Environmental Protection Agency. I. Title.
HC110.E5F55 1995
363.7'00973—dc20

94-28832
CIP

Printed in the United States of America

08 07 06 05 04 03 02 01 00
9 8 7 6 5 4 3

The paper used in this publication is both acid-free and totally chlorine-free (TCF). It meets the minimum requirements of ANSI/NISO Z39.48-1992 (R 1997) (*Permanence of Paper*). ♾

To the memory of Joseph Fiorino, Jr.

Contents

Figures

Tables

Preface

This project originated with a simple question: What does a person need to know to begin to understand how environmental policy is made at the national level in the United States? Like many questions, this one raised many others: What are the critical subjects to address in the space of one short volume? Should the discussion focus on policy processes, the substance of problems and policies themselves, or the analytical frameworks in which policy decisions are made? How much emphasis should be given to the role of institutions in policy making, as opposed to the perceptions of problems or the regulations and other tools devised to solve them? How essential is discussion of international issues to an analysis of domestic policy making? How important is knowing how policy is made now for understanding how policy should be made in the future?

Writing this book is rather like making environmental policy. It is necessary to make difficult choices under situations of constraint. Just as we as a society cannot invest in every problem coming up on our collective radar screen, so the author of an introduction to as broad and diverse a topic as environmental policy cannot deal with all issues and explore every point of view. One has to set priorities and make choices. What follows reveals my choices and priorities.

The hardest choices involved decisions about what to include and what not to include. That this is a book written by a political scientist oriented toward policy settled many of the questions about coverage and emphasis. Decision making and institutions inevitably play a central role in the analysis. And yet economic and scientific issues and methods are essential to a discussion of policy making, as is the array of policy tools for translating the intentions of policy makers—legislators, political executives, judges, and agency leaders—into action. I decided to organize the book around four topics—institutions, analyses, problems, and strategies—presented in five chapters. The introductory chapter presents more on the organization and the theoretical basis for this approach. The concluding chapter looks toward the future.

Here are some of the other choices I made. The interdependence among domestic and international issues made the latter a necessary topic for discussion. At the same time, a focus on domestic policy making required that the international issues support but not take over the analysis. The U.S. Environmental Protection Agency's (EPA's) internal structure and politics affect policy in many ways, so it is prominent in the discussion. Several volumes could be devoted to describing the complex legal framework under which EPA operates. I give an overview of the key statutes to present the variations in approaches and policy goals that the laws incorporate. Risk and economic analyses are central to policy making and a source of controversy, so a chapter is devoted to describing how the analyses are done, the assumptions and uncertainties behind them, and the limitations on their use. The strategies chapter includes a look at command-and-control, or direct, regulation as well as uses of economic incentives and other innovative approaches. Although most of the analysis focuses on how policy is made now, the final chapter looks at trends that will shape policy in years to come.

In addition to decisions about what topics to present and at what level of detail, there were choices to make about several matters of interpretation. The most important of these was the overall picture of policy making in this country. Writing always involves an exercise in constructing a subjective reality, especially with something as abstract as public policy making. The reader relies on the author to paint a picture that makes sense, to provide a view of institutions and people as engaging in something more or less defensible from society's point of view. Is environmental policy made reasonably well in the United States, by capable people working as part of well-designed institutions in the best interests of society as a whole? Or is policy making full of flaws, done by people who should be doing better, with results that are hard to defend from any socially responsible point of view?

My own answer, my interpretation of policy making that defines the presentation here, goes like this: Environmental policy making in this country is a necessary and socially desirable enterprise. Government intervention to deal with environmental problems was an appropriate response to a number of threats to our quality of life and our natural resources. After years of modest experiments with an approach defined by a low federal profile and state discretion, Congress led a major expansion in the federal role in the early 1970s. The result has been an elaborate, ambitious, but flawed set of institutional responses and policy designs for remedying past damages to the environment and reducing future ones. The accomplishments are impressive: a range of laws; a significant national investment, both by the public and the private sectors; a list of improvements in environmental quality, or at least of conditions that might have gotten worse but have not; an array of institutions at many levels of government; a body of regulations, standards, and policy guidelines that other countries often emulate; a corps of professionals across a range of disciplines, which greatly increases the national capacity to identify and respond to problems.

As one would expect from an enterprise of this scope, the flaws are also impressive: a legal structure that breaks issues down into artificially small pieces; a pattern of responding in incremental steps to problems as they become apparent and draw public attention but rarely before; a fragmentation in national institutions; reliance on command-and-control regulation to the near-exclusion of other mechanisms; too much money spent on some problems and not enough on others; an inability to decide what the relationship should be among economic and environmental goals or to recognize that they complement as well as conflict with each other.

My view is that with the exception of the early 1980s, policy has been made and institutions led by people who, for the most part, were capable—in some cases, impressively so—and had a sincere interest in protecting the environment and reconciling environmental with other social goals. Many of these people faced the constraints in laws, institutions, knowledge, and perspectives that are discussed throughout this book. In some cases they tried to overcome these constraints; in others they learned to live with them. Whatever the limits of current approaches, they constitute an impressive set of accomplishments in a complex and controversial area of policy.

I should add one final note regarding proposals to elevate EPA to the status of a cabinet department. In May 1993, the U.S. Senate passed such a bill, which would create the U.S. Department of Environmental

Protection. The bill had been under consideration in Congress near the end of President George Bush's administration and was strongly endorsed by President Bill Clinton early in his term of office. As of May 1994, however, the bill had not yet been reported out of committee in the House of Representatives, because of disagreement over amendments that would require the department to conduct risk assessments and cost-benefit analyses for many of its regulations. The likelihood of passage was uncertain as of May 1994. If the bill were to pass, it would, in addition to changing the name of EPA to the Department of Environmental Protection, elevate the EPA administrator to the status of a cabinet secretary, redesignate assistant administrators as assistant secretaries, and perhaps require certain analyses or findings as part of the regulatory decision process. It would not, however, change EPA's authorities or responsibilities from what they are under current laws or modify its organization or functions in any fundamental way from what is presented in this book.

Acknowledgments

This book is the product of years of thinking, talking, and writing about environmental protection from a policy point of view. Many people have helped make it possible. Tillie Fiorino's help on all of those papers long ago and her support and encouragement made it possible for me to write this book. I want to thank Larry Esterly of Youngstown State University for first teaching me about writing and about political science. Elizabeth Knoll of the University of California Press made the phone call that got me thinking about writing a book of this kind. Her advice and encouragement kept me on track throughout the project. Michelle Bonnice expertly led me and the manuscript through the several stages in the publication process. I am indebted to Sheila Berg for her very sure-handed and skillful editing of the final manuscript. Susan Hadden and Michael Kraft gave valuable comments on an early draft; Susan's detailed comments on the manuscript were especially helpful. I want to thank the students in my American University graduate seminar, "Environmental Policy and Politics," for a last chance to work through the presentation.

My most important debt is to Joanne Fiorino. Her careful and expert review of the manuscript and her many insights made this a far better book than it would have been without her help. She has provided support and encouragement in too many ways to mention. My sons, Matthew and Jacob, provided a stimulating environment for writing and thinking about many issues that affect their futures.

Abbreviations

AA	Assistant Administrators
APA	Administrative Procedure Act
CAA	Clean Air Act
CAAA	Clean Air Act Amendments of 1990
CWA	Clean Water Act
CERCLA	Comprehensive Environmental Response, Compensation, and Liability Act
CPSC	Consumer Product Safety Commission
DOE	U.S. Department of Energy
EIS	Environmental Impact Statement
EPA	U.S. Environmental Protection Agency
FDA	Food and Drug Administration
FFDCA	Federal Food, Drug, and Cosmetic Act
FIFRA	Federal Insecticides, Fungicides, and Rodenticides Act
FWPCA	Federal Water Pollution Control Act
GDP	Gross Domestic Product
GNP	Gross National Product
HSWA	Hazardous and Solid Waste Amendments
HUD	U.S. Department of Housing and Urban Development
MACT	Maximum Available Control Technology
NAAQS	National Ambient Air Quality Standards

NAS	National Academy of Sciences
NEPA	National Environmental Policy Act
NOAA	National Oceanic and Atmospheric Administration
NRDC	Natural Resources Defense Council
NSPS	New Source Performance Standards
OECD	Organization for Economic Cooperation and Development
OIRA	Office of Information and Regulatory Affairs
OMB	Office of Management and Budget
OSHA	Occupational Safety and Health Administration
RCRA	Resource Conservation and Recovery Act
RIA	Regulatory Impact Analysis
SAB	Science Advisory Board
SARA	Superfund Amendments and Reauthorization Act
SDWA	Safe Drinking Water Act
SIP	State Implementation Plan
TRI	Toxics Release Inventory
TSCA	Toxic Substances Control Act
UNCED	United Nations Conference on the Environment and Development
UNEP	United Nations Environment Programme
USDA	U.S. Department of Agriculture
VOCs	Volatile Organic Compounds

CHAPTER ONE

Challenges

An excellent introduction to environmental policy is Dr. Seuss's *The Lorax,* the story of a prospering ecosystem ruined by unbridled industrial development. At the center of the ecosystem is the Truffula Tree, which sustains thriving communities of Brown Bar-ba-loots, Humming-Fish, and Swomee-Swans. The trouble starts when an unscrupulous developer, the Once-ler, moves in, recognizes the tantalizing commercial potential of the brightly colored Truffula tufts, and starts cutting down trees to produce thneeds. The thneeds sell as fast as they are made, which only increases the Once-ler's greed and the rate of destruction of the trees. As the Truffula Trees go, so do the other forms of life that depend on it. With not enough Truffula fruit to go around, the Brown Bar-ba-loots are forced to migrate elsewhere. Choked by "smogulous smoke," the Swomee-Swans soon follow. Finally the Humming-Fish leave too; discharges from the thneed plant (mainly Gluppity-Glupp) into their pond are gumming up their gills.

The one lonely voice against this exploitation is the Lorax, who speaks "for the trees . . . for the trees have no tongues." But the pleas of the Lorax go unheeded. As the Once-ler sees it, "business is business! And business must grow." The exploitation goes on, until the once-thriving, Truffula-based ecosystem has turned into a treeless, polluted landscape that can neither sustain life nor repair itself. Finally, even the Lorax departs. In the end, one Truffula seed remains as the last hope of salvation.[1]

The story of the Lorax and the Truffula Trees gives us an elegant parable of environmental devastation. It describes what could have happened here, what did happen in parts of Eastern Europe and the former Soviet Union, and what yet may happen in many nations still in the early stages of industrialization. The Lorax spoke not only for the trees but for future generations who could have enjoyed them and the life they sustained. But the Lorax was only a voice, lacking inducements or regulatory powers. And so, in the end, the resource was lost, almost irretrievably.

It did not have to be that way. There were alternative paths for developing the Truffula resource without destroying it. The Truffula ecosystem could have been protected entirely. The tree cutting could have been controlled to keep it sustainable. The thneed plant could have been subjected to strict limits on its air and water emissions. The Once-ler might have been persuaded to adopt a more aggressive policy of preventing pollution at the source and matching tree losses with a replacement or restoration program. As for the Lorax, its heart was in the right place, but it lacked a means of turning good intentions into results. The Lorax needed an environmental policy: institutions, strategies, rules, and methods for making choices and carrying them out.

The United States does have an environmental policy, and that policy is the subject of this book. Our problems are surely more varied and complex than those of the Lorax, but the fundamental issues and choices are the same. We worry about how decisions made today limit prospects for the future. We look to prevent selfish interests from achieving their own ends at the expense of society. We try to balance economic growth against its effects on health and the environment. The issues are more complex, the threats to the environment are more numerous, and the stakes are higher, but the parable of the Lorax and the Once-ler is as valid for contemporary American society as it is for the land of the Truffula Trees.

The analysis in this book will take us into several fields. It takes us into politics, because environmental policy is made in an intensely political atmosphere where values and interests often collide. It takes us into science, which enables us to understand problems and attempt to solve them. It takes us into the field of ethics, because few areas of policy present more difficult choices: how to preserve shared resources, how to distribute costs and benefits, how this generation's actions will affect future ones. It also takes us into economics, because a society's choices about the environment relate directly to how it produces, consumes, and preserves its resources. Psychology, law, sociology, and other fields also come into play in our discussion of this eclectic and interdisciplinary area of public policy.

This chapter gives an overview of the chapters that follow. First, however, I want to do two things: (1) introduce the key institutional challenges that environmental policy makers will face in coming decades, and (2) look briefly at various models of public policy and how they influenced the approach in this book.

INSTITUTIONAL CHALLENGES IN ENVIRONMENTAL POLICY

The people who make, manage, or study environmental policy in the coming decades will face a variety of challenges. There will be scientific challenges as we try to understand problems and their causes, explain relationships between exposures and effects, or predict long-term changes in climate or atmospheric chemistry. There will be legal challenges in our efforts to observe the demands of due process, maintain compliance without infringing on individual rights, meet burdens of proof while not paralyzing programs, and cope with the diverse obligations of our laws. There surely will be political challenges—in the need to set priorities, to educate and lead without dictating solutions, to involve the public in complex, technical decisions. And there will be economic challenges as we struggle to adapt contemporary economic systems to the needs and limits of the national and global environment.

Among the challenges, five stand out. These are not the five most pressing environmental problems but the institutional challenges that will determine how well we will define, compare, and resolve problems. If we can meet them, the solutions are more likely to follow.

SETTING THE POLICY AGENDA

A reality for environmental policy makers is that there are more problems demanding attention than there is money, people, knowledge, or political will to solve. American society could double the resources devoted to environmental protection and still leave much of the job undone. The challenge is to focus resources on the problems that most deserve them, to determine priorities by setting the policy agenda.[2]

A recent administrator of the U.S. Environmental Protection Agency (EPA) once compared the nation's approach to environmental priorities to a video game called "Space Invaders." "In that game," he observed, "whenever you see an enemy ship on the screen, you blast at it with both barrels—typically missing the target at least as often as you hit it."[3] The course of the nation's policy over the last two decades, he said, had been much like this game. "Every time we saw a blip on the radar screen, we unleashed an arsenal of control measures to eliminate it." The nation responded to problems piecemeal (which may have been the only politically feasible way to respond at the time), without an overall sense of their relative importance. The result is an elaborate set of programs addressing problems in the order in which they came up on the policy agenda, not necessarily in the order of their importance to ecology or health.

Are these the right problems, given our limited resources? How do we decide in what order and at what level of effort to take on new problems as they come up on our policy radar screens? Should hazardous waste consume a big share of the environmental budget, while indoor air pollution receives almost none? Should protecting the Great Lakes rank above cleaning up Chesapeake Bay, or either of these above removing asbestos from schools or lead paint from public housing? Should we allow the crisis of the moment—pesticides used on apples or medical waste washing up on beaches—to distract us from other issues on our agenda?

Charles Jones suggests three patterns of agenda setting: one in which government takes a relatively passive role and reacts to the play of private interests; a second in which government defines a process and encourages private interests to participate in setting priorities; and a third in which government plays an active role in defining problems and setting goals. Under the third, institutions "systematically review societal events for their effects and set an agenda of government actions."[4] In this book, I advocate a role for environmental agencies more like the third than the first two. Agencies should use their knowledge of problems to shape as well as respond to the policy agenda.

MAINTAINING DEMOCRATIC VALUES

With all the concern about the fate of pollutants, control technologies, sampling and measurement, toxicological models, and other technical aspects of the field, it is easy to forget that making environmental policy is above all else about government. I once began an essay on risk with the question, "Can a technological society remain a democratic one?"[5] There is no easy answer. The emergence of environmental problems is linked to the rise of technology in modern society. Technology has had at least two effects on politics. One is a growth of knowledge in areas that are not easily accessible to people who lack specialized training in scientific or technical fields. Later on, the discussion of risk analysis illustrates this point. The other, related effect is the influence of technical and administrative elites whose knowledge, expertise, and ideas frame the issues and structure the choices government institutions make. Technological experts and elites necessarily play a central role in governmental and especially environmental decision making. This need to rely on technical experts is one of many trends that threaten to take policy choices away from ordinary citizens and place them more in the hands of technical and administrative elites.

This is not just a theoretical issue. American political and social institutions in general have undergone a crisis in public confidence over the last thirty years.[6] There has been a steady decline in the legitimacy of the institutions that make policy decisions, environmental and otherwise, public and private. The technical complexity of environmental decisions and the need to rely on experts only makes the job of sustaining democratic values more daunting.[7] Siting municipal waste incinerators, locating disposal sites for radioactive waste, determining acceptable levels for exposures to toxic chemicals—all of these are choices that require us to reconcile technical and democratic values in our decision making.

We can see the practical effects of this challenge to democratic values in the so-called NIMBY (not-in-my-backyard) syndrome. The American political system provides several avenues for groups in society to express opposition to environmental decisions. At the local level, this may take the form of opposition to the unwanted necessities of technological society: nuclear plants, waste incinerators, municipal landfills. A combination of greater sensitivity to the effects of hazards and a loss of confidence in institutions and leaders creates a dilemma for policy makers. As a result, the conventional policy process often "has been rendered incapable of effectively balancing needs for growth, development, and facility siting with those of health and environmental protection for current and future generations."[8]

Another practical effect of this challenge to democratic values is a greater use of statewide voter initiatives to make critical choices about health and environmental policy. California is the state that has turned most often to initiatives, which allow voters to express their dissatisfaction with legislatures and with conventional political processes.[9]

California's Proposition 65, the best known of these environmental initiatives, established a range of public notification and other requirements for a list of toxic chemicals. A more sweeping environmental initiative (known as "Big Green") failed a few years later in California, as did broad environmental initiatives in Ohio and Massachusetts in 1992. But the likelihood of citizens choosing to make policy through the direct democracy of voter initiatives remains high in many states. They present citizens with a mechanism to play a more direct and determinative role in environmental decisions. At the same time, they reflect dissatisfaction with environmental politics-as-usual.

A characteristic of American political culture is that people value the opportunity to take part in decisions that affect them.[10] Add to this the traditional American skepticism toward technical experts and centralized power, and the challenge to policy makers is clear.

USING SOCIAL RESOURCES EFFICIENTLY

"Nothing in life is free" is as true in environmental policy as it is in anything else. It is a fact of life that resources dedicated to one set of social goals are unavailable for others. What we spend as a society to reduce ozone levels, remove lead from drinking water, or restore wetlands cannot be used to expand prenatal health care, improve public education, build aircraft carriers, or expand the national park system. Within environmental programs, resources that are devoted to air toxics, asbestos removal, or sewage treatment cannot be used to save wetlands, invest in new air pollution control technologies, or remove lead from paint in old houses. Economists think in terms of opportunity costs, in which we evaluate the worth of one set of expenditures against others that are given up.[11]

A premise in this book is that policy makers in any field—whether it is environment, health, defense, or education—should use society's resources efficiently. This does not require a strict cost-benefit accounting for all policy decisions, but it does suggest that we should know what we are getting in return for our environmental investments and that we should design ways to use scarce resources wisely. We will see that economic analysis and its role in environmental policy are controversial issues. Some people argue that environmental risks should be kept as close to zero as is technically possible, whatever the cost. Many of the provisions in the national environmental laws reflect this point of view, especially with regard to hazardous waste. But it is hard to ignore the fact that money spent for one thing cannot be spent for another.

When we make a choice as a society to require the use of a particular technology or achieve a given environmental quality goal, we have decided implicitly to devote resources to that end over others. Such choices are better made explicitly, with some awareness of costs and benefits. With environmental protection taking a noticeable share of society's resources, with a variety of other social needs calling for attention, and with the political necessity to keep the U.S. economy competitive, policy makers should analyze the consequences of their choices and use resources well.

ADAPTING INSTITUTIONS

Modern societies organize themselves to make collective decisions through political and social institutions. Success at dealing with environmental problems depends on the strength and adaptability of institutions and the relationships among them.

Each of the challenges described in this chapter involves institutional change. But some adaptations are important enough to consider on their own. Three kinds of institutional adaptation are especially important. First, policy makers will have to integrate environmental programs (e.g., air or water) and policy sectors (e.g., environment, energy, or agriculture) more effectively than they have in the past. A weakness in the current process for making environmental policy is that problem definition, analysis, and decision making are fragmented. Consider the many signs of fragmentation: EPA is less an integrated environmental management agency than a holding company that is responsible for implementing more than a dozen major laws. Each of our major environmental statutes defines problems and sets standards for action differently from the others. Problems tend to be broken into artificially small pieces, with air, water, and waste issues often considered separately from one another.[12] Beyond EPA, we need to better integrate the environmental sector with other policy sectors—such as energy, agriculture, land management, transportation, and foreign policy—where the implications for environmental quality are huge.

Second, the capacities for addressing international problems will have to be enhanced. A hard fact of life for environmental managers is that problems do not follow political boundaries. Some problems are bilateral, like cross-border air pollution or pollution flowing into San

Diego Harbor from the Tijuana River, and require joint action by two nations. Others, such as the quality of the Mediterranean or Caribbean Sea, are regional and require action by several nations. Still other problems, such as atmospheric warming or stratospheric ozone depletion, are truly global; to solve them, international institutions will have to expand their capabilities and authorities. The environment will have to be more prominent on foreign, trade, and economic agendas.

A third kind of adaptation will be in the way public and private institutions relate to one another. It is an understatement to say that relations between environmental agencies and regulated industries have been difficult for most of the last twenty years. Relations could fairly be described as adversarial, characterized by conflict and distrust. There was an "us" versus "them" mentality—industry resisting regulatory controls at every step, agencies fighting their way through a tangle of lawsuits and political opposition. Many observers contrast the adversarial features of policy making in the United States with the more cooperative, consensual policy making in such nations as Britain or Sweden.[13] There are recent signs of public–private cooperation in policy making, pollution prevention, and financing, but there still is room for progress in relationships between the public and private sectors.

MEASURING AND EVALUATING PROGRESS

Common sense dictates that in any area of policy we would have a reliable way of knowing how we are doing, of assessing trends in the extent and severity of problems and successes or failures in solving them. Policy makers need a steady flow of information to enable them to set priorities, design strategies, and make policy choices. They need "indicators" to define acceptable measures of progress.[14]

Such indicators are common in many areas of policy. Most of us are familiar with the quarterly reports on unemployment issued by the Bureau of Labor Statistics. Economic policy relies heavily on indicators that are followed widely and used by the private and public sectors to make decisions—the consumer price index, the Dow-Jones industrial average and other market indicators, key interest rates, indexes of industrial production, and so on. In environmental policy, we lack indicators as widely accepted and relied on as these.

Policy makers have had some success at knowing in what direction the country is headed when it comes to emission levels or the environmental conditions in specific areas. For example, emissions of the principal pollutants regulated under the Clean Air Act (CAA) have declined since 1970. Still, many people live in areas where levels of these same pollutants in the ambient air exceed federal, health-based standards. For water quality, there have been reductions in pollution from many large industrial plants, but these have been offset by increases in pollutants that run off from nonpoint sources like agricultural fields and construction sites. So there is good and bad news; there are signs of progress and reasons for concern.

Yet what we know falls short of giving us reliable sets of environmental indicators. There are two problems with what we currently know about environmental trends. First, the data are uneven—over time, across problems, and by geographic area. For example, national trends in air quality tend to be documented more fully than trends in water quality, and information on Chesapeake Bay is more complete than is information on the Great Lakes. Second, it is difficult to link the measures we have to program performance and environmental results. What effect do controls on nonpoint source water pollution have on sensitive ecosystems? How fully do technical standards for hazardous waste disposal sites protect nearby groundwater? How have trends in ozone levels in U.S. cities affected the lung capacity of children growing up in those cities?

The problem of determining cause and effect is not limited to environmental policy. Linking cause and effect as a basis for evaluating the effectiveness of government policies is one of the oldest and most intractable tasks of public policy. What have been the effects of the Head Start program on the educational achievement of inner-city children? What effect will alternative financing schemes have on the costs and quality of the nation's health care? In part because of this, we face formidable technical and political obstacles in devising sound indicators for environmental quality. Yet it is hard to know how well we are doing if we lack measures of how far we have come and projections of where we are going.

MODELS OF PUBLIC POLICY

One of my objectives in this book is to examine environmental policy within the broad context of the study of public policy. To make a link with the public policy literature, I draw on several models of or approaches to public policy. A model is a simplified description of reality that can help explain an object or phenomenon. Administrative theory offers several kinds of models for understanding policy making.[15] We will look at them briefly in what is my only discussion of theory.

I begin with a look at four models that have influenced the study of public policy over the years: institutional, systems, group process, and net-benefits. Each offers useful insights into the study of policy making; at the same time, each has certain limitations that make it unsatisfactory for explaining policy making on its own. Next, I examine two concepts that offer a foundation for another approach, based on Herbert Simon's concept of "bounded rationality" and Charles Lindblom's concept of "incrementalism." I then outline another approach to the study of public policy. In this approach, the political scientist John Kingdon built on the "garbage can model" of organizational choice to develop yet another model for public policy, one that shapes the approach taken in this book.

INSTITUTIONAL, SYSTEMS, GROUP PROCESS, AND NET-BENEFIT MODELS

The oldest model for studying public policy making is the institutional model, which describes and analyzes institutions, laws, and procedures. An institutional approach is mostly descriptive and limits itself to the formal, legal influences on policy—statutes, court cases, and administrative organization or procedure. A pure institutional analyst would see policy making as entirely the product of these formal arrangements.

A second approach is the systems model. Systems theory emerged early in this century and is founded on an analogy between biological and social phenomena. Political and social life is seen as a series of interrelated systems, each of which must cope with and adapt to changes in its environment if it is to survive. In administrative theory, the most commonly studied system is the organization, such as a public agency. Systems theory attempts to explain an organization's behavior by analyzing the inputs into its decision making, the products or outputs that emerge from it, and the processes that convert the former into the latter. Organizations monitor events in their external environment as a source of feedback that enables them to adapt to change and maintain equilibrium. For our purposes, systems theory offers a useful picture of how organizations interact with forces in their environments, whether they are forces such as interest groups and courts or broader demographic and economic changes. It provides a dynamic view of the behavior of organizations as they adapt to external forces.

A third approach is the group process model, which dominated political science for years after World War II.[16] The unit of analysis is interest groups in society; policy is the outcome of competition for influence among them. The relative power of interest groups determines the substance of policy and the values that government promotes. For example, a critique of economic regulatory policy from the point of view of the group process approach is that agencies proceed through life cycles that eventually make them vulnerable to "capture" by the very industries they were set up to regulate.[17] The result is regulatory policies that promote the narrow interests of the regulated industry over broader public interests. In a well-known critique of the group process model, Theodore Lowi argues that it justifies a conception of government as little more than a passive arbiter of the preferences of the dominant interests in American society. He is critical of the assumption that the policies that are most acceptable to organized and influential interest groups in society are the best policies overall for government institutions to adopt.[18]

A fourth approach is the net-benefits model. It presents a counterpoint to the group process theorists. The group process approach is the product of political science, whereas this one is more the product of economics.[19] Group process theorists see administrators as referees who reconcile the diverse interests in society to achieve results that are acceptable politically to the groups that represent those interests. In contrast, under the net-benefits model, administrators are seen more as analysts who should make decisions that offer the greatest net benefit or "utility" to society. In the group process model, the test of a "good" decision is whether it is acceptable to affected groups, whatever the economic or other consequences. In the net-benefits model, a decision is considered to be "good" when it meets the criterion of economic efficiency, defined ideally as the state in which no party can improve its position without worsening that of another.[20]

To make decisions that maximize net social benefits, one follows a rational process: define policy options, quantify the likely effects of each, compare them with a set of objectives, and then select the one with the best ratio of benefits to costs. This approach to regulatory decisions became influential in the 1980s under the Reagan and Bush administrations, which advocated cost-benefit analysis as a basis for making environmental decisions.[21]

Each of these approaches helps to explain how policy is made. Clearly, laws, institutions, and procedures shape the process and substance of policy. It would be difficult to make sense of air quality programs without understanding the structure of the Clean Air Act, the relations among national and state agencies, or the legal and administrative process for setting air quality standards. The systems model places the policy making organization (EPA, for example) in the context of a larger environment and sheds light on the interactions between the organization and its environment. It also helps us understand policy as a dynamic process, rather than as the more static one implied by the institutional model.

The group process model clearly has relevance in a society like ours in which political power is decentralized and policy is largely shaped by bargaining and negotiation among interest groups. Whether the subject is the Chemical Manufacturers Association, the American Petroleum Institute, or the Environmental Defense Fund, the influence of interest groups on policy is undeniable. And yet the net-benefits model has had a major effect on environmental policy during the last two decades. The ideal of decisions that conform to cost-benefit goals is unlikely ever to be adopted in practice. But this model defines a set of analytical tools and principles that have affected the U.S. approach to environmental issues. Much of our discussion of analyses testifies to the influence of the net-benefits model. Still, the group process model explains how many policy decisions are made. Differences between the two models underlie much of the current policy debate.

Each model also has limitations. The institutional model does not account well for informal relationships or patterns of behavior. A systems approach sheds light on how organizations cope with and adapt to their environments but is less useful for understanding the internal processes that determine policy. The group process model takes us outside of formal government processes and depicts the interplay of interest groups. It is less helpful in explaining internal, bureaucratic influences on decisions. The net-benefits model has affected the outcome in a number of key environmental decisions in recent years. But so many decisions do not follow its analytical criteria that it clearly offers only a partial explanation. Despite the best efforts of economists and other proponents of the net-benefits approach, most environmental decisions do not appear to maximize net benefits, at least as they are defined in the model. In sum, each of these models tells us something important about policy making, yet all fall short in some way in enabling us to understand environmental policy making.

BOUNDED RATIONALITY AND INCREMENTALISM

To broaden our search for models and concepts that may prove useful in understanding how environmental policy is made, we can look at the work of Herbert Simon and Charles Lindblom. They laid a foundation for the garbage can model, especially in their criticisms of the "rational" models of policy making that many earlier theorists had espoused. Before turning to their work, I will examine the concept of rationality in policy making.

The search for "rationality" has been an important theme in administrative theory and in the study of public policy as well. For our purposes, rationality is defined simply as a conscious decision to make the most of the available resources (including not only money but talent, time, political capital, and so on) to achieve whatever it is one sets out to accomplish. This is a far less rigorous definition than that used by most theorists, but this book takes a more practical look at policy making than most.

In administrative theory, however, the concept of rationality and a rational model imply something more ambitious. As the name suggests, a rational approach to making public policy assumes a decision process in which goals are clear and agreed upon, policy options and criteria for evaluating them are defined, and information about the consequences of options is complete. In a rational process, decisions are made in a linear, sequential way by fully informed people. If the people making policy decisions are doing their jobs, they make decisions that "optimize" their own as well as society's interests by deriving maximum benefits from whatever choices they make.

It takes very little experience in large organizations to realize that they do not work in this way. In making a decision about how to respond to an international crisis, for example, the rational decision maker would have nearly complete control over the situation, an accurate estimate of goals and constraints, fairly complete and reliable information, and a clear sense of what the national interest requires in the situation. Yet even in such a crisis, decision makers are not (and cannot be) as rational as we would like to think they are.

In a study of the Cuban missile crisis and how the Kennedy

administration handled it, for example, Graham Allison suggested two alternative models of decision making in place of a rational model. He used a bureaucratic politics model to explain the U.S. response to the discovery of missiles in Cuba in terms of the competition for influence among various bureaucratic interests in the U.S. government. He used an organizational process model to explain the U.S. response as a product of the "standard operating procedures" of organizations within the State Department, the National Security Council, the White House staff, the Defense Department, and elsewhere.[22] Each of these alternatives offered a more convincing explanation of what happened than did the rational model alone. This does not mean that the Kennedy administration was irrational in its response to the missile crisis. It does support the notion, however, that the conventional ideal of rationality in public policy is usually unachievable.

In most large, complex organizations (or governments, for that matter), the situation looks more like this: goals are ambiguous or conflicting; human cognitive skills are limited; time is in short supply; and policy options are fluid or poorly defined. Simon recognized this in his criticism of the rational approach.[23] He argued that we should accept certain realities about limits to rationality in organizations. He used the terms "satisficing" and "bounded rationality" to describe how decisions are made in practice, rather than in a theorist's ideal conception. Simon argued that decision makers do not optimize the achievement of goals when they make choices, as a rational approach suggests. Instead, they "satisfice." They select the first acceptable policy alternative that comes along. The concept of bounded rationality recognizes that although people may aspire to be rational, they are "bounded" by the limits in resources, time, information, and human cognition. The best they achieve is a bounded rationality.

Simon offered one set of criticisms of the rational approach. Another set of criticisms is Lindblom's incrementalism. Again, the thrust is that a rational model of decision making does not recognize the limits in human and organizational capacities. As an alternative to a rational model, Lindblom proposed incrementalism both as a description of how policy is made and as a prescription for how it should be made in practice.[24] In the incremental model, policy is made in small steps, at the margins of choice, through "successive limited comparisons." Decisions one year rely on those decisions that were made in the past. Options and criteria change as the available information changes and there is a need to develop consensus among affected parties. Policy tends to happen as much as to be decided on.

The incremental approach offers several advantages over the rational one, Lindblom argued: policy mistakes are more easily corrected, because choices at the margin can be reversed; goals emerge as choices are made, so agreement on the goals can come gradually; there is more stability to policy when past experience sets the basis for new actions; piecemeal decision strategies are better for coping with limits in time and information than are radical ones. In place of the rational model and its fully informed, interest-maximizing decision process, Lindblom gave us a model of decisions made through a process of "muddling through." He proposed this as a more realistic and, in the end, a more desirable way of making policy decisions.

The concepts of bounded rationality and incrementalism have influenced the approach to environmental policy taken in this book. The notion that policy makers can meet the noble aspirations of the rational model is inconsistent with reality. Environmental policy making is full of uncertainty, of conflicts over goals, of multiple sources of decision authority and control—all the characteristics of an incremental policy system. Yet policy makers can and should aspire to be more rational in the way they use available resources to achieve whatever it is they and American society have set out to accomplish. Our goal, then, should be a more realizable one: to achieve a bounded rationality in environmental policy making.

FROM THE GARBAGE CAN TO THE STREAMS AND WINDOWS MODELS

All of this brings us to the garbage can model, which relies on Simon's and Lindblom's work. First applied to universities, it describes government bureaucracies as well. Both are organized anarchies rather than the hierarchical and orderly organizations portrayed in the rational approach, in which decisions are made in a linear, sequential, and clearly defined process. In an organized anarchy, goals are unclear or even in conflict. Participation in decision making is unpredictable and fluid. An agency is more a loose collection of ideas and proposals than a well-ordered structure. Information comes into play at multiple points in the process of making decisions and is interpreted in various ways. In the garbage can model, an organization is a "collection of choices looking for problems, issues and feelings looking for decision situations in which they might be aired, solutions looking for issues to which they might be the answer, and decision makers looking for work."[25]

In organized anarchies, decisions are made in four "streams": problems, solutions, participants, and choice opportunities. Problems arise and disappear, change shape or significance, and are combined or separated over time. Policy makers may draw their solutions to problems from a standard tool kit, or entrepreneurs may advocate and win approval for innovative solutions that define new ways of responding to problems on their agenda. Participants move in and out of situations in which choices are made, termed "choice opportunities" in the model, as they look for chances to promote their ideas or themselves. These streams rarely connect; when they do, their "coupling," as Kingdon calls it, leads to major policy decisions.

Kingdon turned the garbage can model of how choices are made in organizations into a description of the broader policy process in the United States. Adapting the model slightly, he sees policy making as made up of three streams. The first, the problem stream, is the process by which conditions or issues come to be defined as problems and thus as a focus of government action. The second is the political stream; events, trends, institutions, and interest groups determine which problems will receive attention on the governmental agenda. The third is the policy stream. This shapes the decision agenda, the list of policy alternatives considered for responding to problems. Ideas, analyses, arguments, and the less visible participants in policy (bureaucrats, congressional staff, think tanks) influence policy the most in this stream. Persuasion and argument are more important here than in the political stream, which is determined more by bargaining and political maneuvering. At some point, the streams come together as policy entrepreneurs take advantage of "windows" of opportunity to change policy.

Of the several models described above, my approach is most consistent with the garbage can and streams and windows approaches. More than others, they convey the dynamism, fluidity, pragmatism, and idiosyncrasy of policy. They influence the approach to institutions, analyses, problems, and strategies used here.

Here, the political stream is represented by the institutions and the people acting within them. My discussion of analyses draws on the concept of the policy stream by examining the roles that risk and economic analysis can play in environmental policy. The concept of the problem stream influenced my analysis of how policy makers define and organize environmental problems, of how they set priorities among them, and of how problem definitions affect policy makers' use of strategies for responding to problems.

Yet the discussion in this book draws on several of the other models as well: the analysis of formal organization and laws relies on the institutional approach; the study of how different interests organize and compete for influence in the policy process reflects the lessons of the group process model; the discussion of analysis and its role in making environmental policy draws on the net-benefits approach; and at many points in the book, I use the concepts of bounded rationality and incrementalism to describe and explain various aspects of policy making. Together, these models give us a foundation for analyzing environmental policy in the United States.

AN OVERVIEW

My discussion of theories ends now. But it will be clear throughout the book that each of the kinds of models discussed in this chapter can help in understanding environmental policy. The plan from here on is as follows.

Chapters 2 and 3 provide a survey of the institutions that make or affect national policy, with EPA as a focus. They examine the statutory framework for environmental policy, the organization and operation of EPA, and the process for developing environmental regulations. But EPA is only one player on a crowded stage. In chapter 3, we turn to the various institutions outside of EPA that determine or try to influence environmental policy in the United States. This survey of the institutional landscape presents a number of plots and subplots: the struggle between the executive and legislative branches of government; the debate over the role of the courts; the search for cooperation and policy integration across agencies and sectors (i.e., energy, agriculture, trade); the search for balance among federal and state authorities; and the response of international institutions to global issues.

Chapter 4 introduces the types of analyses that are used to make environmental policy. A key concept is risk. Risk defines the nature and contours of environmental problems. When we speak of an environmental problem, we generally speak in terms of risk: the risks of exposure to toxic chemicals such as benzene and asbestos; the risks of new technologies such as bioengineering or nuclear power; the risks of phenomena such as global warming or losses in biological diversity; or the risks of conditions such as elevated levels of carbon monoxide in urban air. Perceptions of risk influence what we view as problems and how we give priority to some over others. During the last decade, risk assessment has become almost routine in environmental policy.

Costs and benefits are two other concepts that are central to policy analysis for setting priorities and deciding when and how to respond to identifiable risks. The terms "costs" and "benefits" also carry a great deal of political baggage in policy debates. Costs are undesirable or negative effects of a condition or an action; benefits are desirable or positive ones. High ozone levels in Los Angeles' ambient air involve many costs, in the form of harm to health and damage to resources. By reducing those ozone levels, people in Los Angeles will enjoy benefits—better health, a more desirable quality of life, less damage to materials. But reducing ozone levels also involves costs—emission controls, lost productivity, altered lifestyles. The analysis of these costs and benefits has influenced how policy makers have responded or propose to respond to environmental problems.

Chapter 5 is about environmental problems. The range and difficulty of the environmental problems facing the world are simply daunting. New problems keep emerging, older ones rarely go away, and even the ones that we think we understand may take on new dimensions. At great expense, this country built treatment plants that remove most contaminants from sewage; it then had to come up with ways to dispose safely of the sludge by-product in which the contaminants reside. Ten years ago, few people worried about radon in homes or the gradual warming of atmospheric temperatures. Now EPA rates the first as a major health threat and the second as a major ecological risk.

This chapter looks at how we define, organize, and classify environmental problems. Think of just some of the problems on the national agenda: contaminated groundwater, particulates in the ambient air, asbestos in school buildings, benzene emissions from petroleum refineries, runoff of fertilizers and pesticides used in farming, climate change, pesticide residues on food, acid rain, pollution of coastal waters, leaks from chemical plants, fish kills in a river. The list goes on. Aside from their sheer number and variety, two aspects of environmental problems stand out. The first is that definitions of problems rely on multiple, often conflicting principles: some focus on contaminants (asbestos), others on sources (refineries or farming), others on phenomena (climate change) or path of exposure (ambient air). The second aspect is that there are so many problems with such different effects that it is difficult to compare and set priorities among them.

Chapter 6 turns to the strategies and policy instruments that policy makers use to respond to problems. Conditions that are seen as problems provoke a response from policy makers. In responding to problems, policy makers draw on a standard set of policy instruments—rules, incentives, information, grants or subsidies, and others. When government institutions respond to a problem, they draw on and group these instruments into more or less coherent strategies. Strategies may emerge full-blown from legislatures or piecemeal from agencies. Some are national in scope, others regional or local. Some strategies rely solely on direct regulation, others on economic incentives. Many have worked; some that we have hardly used deserve more attention.

Chapter 7 revisits the main themes, considers some trends in environmental policy, and presents two visions of the future—one optimistic, the other pessimistic. It concludes with observations on the concept of rationality in policy making.

A NOTE ON KEY TERMS USED IN THIS BOOK

Before getting into the discussion of institutions, I want to clarify some terms that come up frequently in discussions of environmental policy. Each is addressed in more detail in later chapters.

Ambient and Emission Standards As will be seen in the discussion of strategies (chap. 6), policy makers distinguish "ambient" standards from "emission" standards. Ambient standards refer to the environmental quality in the surrounding air or in a body of water. They measure levels of concentration of pollutants (such as parts per million). Emission standards describe the limits on the amounts of pollution coming from a specific source, for example, a smokestack or a water discharge pipe. Of course, one reflects the other; the greater the emissions from air pollution sources in an area, the worse we can expect the ambient air to be. But so many factors affect the relationship between emissions and ambient air quality—among them, weather conditions, patterns in emissions, topography—that ambient conditions may vary, even with the same level of emissions. Another point is worth making here: people in the environmental field usually use the term "emissions" to describe pollution from air sources and "effluent" to describe pollution from water sources (which are often termed "dischargers").

Point and Nonpoint Sources The identifiable sources of air emissions and water effluents are usually described as "point" sources. They were the focus of environmental policy until recently. The more scattered, harder to identify sources of pollution are termed "nonpoint" sources. These include pesticides and fertilizer that run off farmland into streams. They also would include the contaminants that flow off a construction site into a nearby stream, if the contaminants are not collected and treated in some way. Because point sources are easier to identify and regulate and often consist of large industrial facilities, they have usually been subjected to controls first. But nonpoint sources account for much of the environmental pollution that remains, especially water pollution, and policy makers have turned their attention more to such sources.

Conventional and Toxic Pollutants There is such a tremendous range of different kinds of environmental pollutants that policy makers have adopted certain distinctions among them. One of the more important ones is the distinction between "conventional" and "toxic" pollutants. Conventional pollutants were the subject of early efforts to reduce pollution. They include common and persistent pollutants that are harmful over time and in quantity but are not necessarily dangerous at typical levels. In the air program, for example, these include such common pollutants as carbon monoxide, sulfur oxide, and small particles. These pollutants pose harm, often serious harm over time or at high levels, but in many cases we have learned to live with them. Toxic pollutants tend to pose more harm, even at small quantities, and the aim of government policy has been to severely limit or even eliminate them. Benzene, mercury, lead, and dioxin are examples of toxics. Policy makers tended to focus more on conventional pollutants in the 1970s, then became more concerned about toxics in air, water, and waste in the 1980s. Both kinds of pollutants are the subject of government policy, but they tend to be treated differently in law and regulation, with tighter controls usually imposed on toxics.

Media Programs and Policy Sectors Government policy has generally been organized to address problems according to the environmental medium through which people and resources are exposed to harm—surface water, drinking water, air, waste, or pesticides. EPA's programs tend to be organized on this basis, so when I refer to environmental "media" or "programs," I am often referring to the same thing. I use the term "sectors" to distinguish among the main areas of public policy; environmental policy is thus a "sector," as are agriculture, energy, trade, transportation, and so on. A theme in environmental policy, and in this book, is the need to integrate policy making better across environmental media and policy sectors.

A glossary of terms frequently used in the field of environmental policy and the study of public policy appears at the end of the book. This may prove especially useful when we turn in the next chapter to one of the more complicated topics in public policy—namely, the statutory framework for environmental protection.

CHAPTER TWO

Institutions I

In a perfectly rational world, a philosopher-king/queen would make all environmental decisions. Toxic chemicals would be managed in the best way possible; if they presented unacceptable health risks, they would be banished. Wetlands would be preserved at a rate that would maximally benefit society in the long run. With budgets for environmental programs limited (as we assume they would be, even for philosopher-kings), there would be one place where all kinds of risks were compared and well-designed strategies devised for dealing with them. Most important, one set of goals would guide policy makers, who would have one source of strategic direction.

But the world of environmental policy is messier and noisier than that. The price of democracy—which more and more of the world has decided is worth paying—is that people and institutions often work at cross-purposes, often toward conflicting ends, with not always satisfying results. Why else have a system in which Congress passes a law directing an agency to set strict regulatory standards, only to allow an agent of the White House to delay their issuance? Why have agencies spend years making a case for banning a chemical, only to have a judge remand it for better justification? And why has Congress divided oversight authority among more than a dozen subcommittees and a score of laws, when one law makes more sense? Institutions have varied perspectives and roles; if they appear to be working at cross-purposes, it is because they often are.

Remember that American politics and policy making represent a series of compromises, a delicate system of checks and balances in which power is no sooner granted to one person or institution than another steps in to limit it. Congress makes laws, as any grade school civics text tells us, but delegates power to agencies such as EPA to carry them out. Agencies make policy under delegated authorities but remain accountable to the courts through judicial appeals and to Congress through legislative oversight. Courts can reject an agency's decision and remand it for further action but depend on an agency to comply with the ruling or on outside parties to challenge its response. The president appoints top policy makers in agencies, may review regulatory decisions, and greatly influences budget allocations but relies on agency staff for technical expertise and to contend with Congress and the courts.

In this and the following chapter, we turn to the several institutions that make and carry out environmental policy. We can start with an observation: if this were a baseball game, it would be a very crowded playing field. The heavy hitters would include Congress, the president, the executive office staff (especially the Office of Management and Budget [OMB]), and the federal courts. EPA would be out there on the mound, in the role of the tired reliever, serving up pitch after pitch, succeeding at times but often being pounded. In the field are environmental groups, interest groups, and trade associations, playing their specialized roles. State agencies and international organizations are also in the game, playing more all the time. Occasionally the benches clear and there is an all-out fight. And there is no manager in the dugout.

This chapter examines the statutory framework for policy making and the history, organization, and operation of EPA. Enabling statutes passed by Congress establish the legal basis for the national environmental programs. In the best tradition of American incremental policy making, EPA's authority is set out, not in one law, but in many, each focused on an environmental medium—air, water, and so on. The focus in these laws is EPA, to which Congress grants authority to develop and oversee the national programs. Rule making—the process by which agencies prepare and issue regulations—is an important form of policy making. It is discussed in detail below. Chapter 3 looks at the heavy hitters—Congress, the executive branch, and the courts—and then at the intergovernmental and nongovernmental influences on policy.

THE STATUTORY FRAMEWORK

We know from Public Administration 101 that administrative agencies derive their authority from their enabling legislation. Some agencies enjoy the relative luxury of having their authority embodied in one, comprehensive, "organic" statute. In an entirely rational world, environmental policy would be governed by one or perhaps a few laws that define agency missions, comprehensively address risk, account for cross-media and interprogram effects, and balance economic with environmental and other goals. It might even bring all natural resource protection programs under one legal umbrella—the organic statute referred to above.

But we live in a less than rational world. The statutory framework is more complicated and less consistent than many would like. Policy is made within a framework of a dozen major laws and several lesser ones.[1] Congress laid the foundations for EPA's programs by passing the National Environmental Policy Act (NEPA) in 1969, the Clean Air Act (CAA) in 1970, and the Federal Water Pollution Control Act (FWPCA) in 1972.[2] The last two defined EPA's early role. Both established a strong federal presence, set ambitious national goals, forced the development of new technologies, and committed the nation to aggressive standard setting and enforcement. They cast a mold that has not changed much in over two decades.

Later in the 1970s came the Toxic Substances Control Act (TSCA) and the Resource Conservation and Recovery Act (RCRA). At the time, the TSCA was the more heralded of the two, but the RCRA has had a far greater effect on the environment and the economy. Concern about inactive waste sites led Congress to pass the Comprehensive Environmental Response, Compensation, and Liability Act (CERCLA), known as the Superfund, in 1980. First enacted in 1947, the Federal Insecticides, Fungicides, and Rodenticides Act (FIFRA) was amended in 1972, just after the authority for administering it moved from the U.S. Department of Agriculture (USDA) to EPA. It was amended again in 1988. Passed in 1974 and amended in 1986, the Safe Drinking Water Act (SDWA) directs EPA to limit contaminants in the drinking water supply.

EPA's legal portfolio does not end here. Several other less sweeping but still significant laws define more specialized roles and authorities, among them, the Marine Protection, Research, and Sanctuaries Act, the Uranium Mill Tailings Control Act, the Medical Waste Tracking Act, the Asbestos Hazard Emergency Response Act, the Pollution Prevention Act, and the Oil Pollution Act.[3] A review of all the statutes that affect EPA and environmental policy making is beyond the scope of this book. I focus on the major programs and statutes governing each, beginning with a survey of the major statutes in five areas—air quality, surface water, hazardous and solid waste, pesticides and toxics, and pollution prevention. These are compared in summary form in table 1.

AIR QUALITY

The Clean Air Act (CAA) of 1970 was not the first federal air quality law. But it differed from previous laws (enacted in 1955, 1963, 1965, and 1967) in important ways. First, it defined national goals in the form of ambient standards (levels of air pollution in the surrounding air) that all parts of the country must meet. Second, it directed EPA to set national standards for controlling emissions of toxic air pollutants. Third, it set limits for emissions from mobile sources (cars and later trucks), as "percentage reductions" from existing levels. Fourth, it gave EPA authority to set national standards for emissions from new sources of pollutants. It also directed the states to prepare State Implementation Plans (SIPs) describing how they would reduce emissions from existing sources in areas (air quality districts that are defined in the act) that exceeded the ambient air quality goals.

A cornerstone of the law was the National Ambient Air Quality Standards (NAAQS), to be set by EPA, based solely on the protection of public health and not on economic factors. The NAAQS reflect EPA's judgment about the maximum concentrations of pollutants that could exist in the ambient air and not harm the health of the most sensitive parts of the population, such as the elderly or people with respiratory conditions. The law further requires that these ambient standards be set to provide an "adequate margin of safety." EPA issued separate NAAQS for several common air pollutants—sulfur dioxide, ozone, particulate matter, carbon monoxide, nitrogen oxide, and lead. Each state had to show that each of its air quality districts had attained the NAAQS or were taking steps through their SIPs to attain them. Any area not meeting the NAAQS was classed as in a state of "nonattainment" for that pollutant.

Much of the rest of the law is aimed at meeting the NAAQS. Congress directed reductions in auto emissions of 90 percent for carbon monoxide and for hydrocarbons and 82 percent for nitrogen oxide. For stationary sources (factories, refineries, utilities, etc.), the law distinguishes new from existing sources. "New" sources are subject to national, uniform New Source Performance Standards (NSPS). The

TABLE 1 U.S. STATUTORY FRAMEWORK (1969–1993)

	Sources/Problems Addressed	*Enacted/ Reauthorized*	*Significant Features*
Air quality	Industrial and mobile sources of air pollutants and air toxics	1970 (CAA) 1977 (Reauth.) 1990 (CAAA)	Combines health-based and technology-based rules. State plans are key. Covers conventional and toxic emissions.
Surface water	Industrial and municipal sources; later "non-point" sources	1972 (FWPCA) 1977 (CWA) 1987 (WQA)	Uses controls on industrial sources and grants for sewage treatment. Relies on some geographic-based programs.
Waste	Current waste generation and disposal (RCRA) Cleanup of abandoned waste sites (CERCLA)	1976 (RCRA) 1980 (CERCLA) 1984 (HSWA) 1986 (SARA)	RCRA/HSWA sets comprehensive rules stressing technology controls. CERCLA cleanup based on "polluter pays" and a trust fund.
Toxics/ Pesticides	Chemicals used as pesticides (FIFRA) Other chemicals posing "unreasonable risk" (TSCA)	1972 (FIFRA amended) 1976 (TSCA) 1988 (FIFRA amended)	Both employ a risk-benefit or cost-benefit test. FIFRA establishes a national registration program.
Examples of other environmental laws	Pollution Prevention Act (PPA) Safe Drinking Water Act (SDWA) Endangered Species Act (ESA) Oil Pollution Act (OPA) National Environmental Policy Act (NEPA)	1990 1986 1973 1990 1969	These address a variety of more specialized problems and issues. EPA is responsible for most but shares responsibility for others (such as NEPA or ESA).

standards are set by industrial category and apply uniformly to all new sources of pollutants covered by the NAAQS. They apply even in areas that have met the NAAQS. The NSPS are technology-based limits; they reflect engineering judgments about what is the best technology available and affordable in a category of industry. The law does not require national emission limits for "existing" sources; these are set by states as necessary to meet or maintain the NAAQS. That is, if a designated area fails to attain the NAAQS and is considered to be in "nonattainment" with the NAAQS, it must set limits on emissions of existing sources or take other measures needed to achieve attainment.

The law also requires EPA to list and regulate several hazardous air pollutants (also known as air toxics), such as benzene and asbestos. These differ from the "conventional" pollutants covered by the NAAQS, because they may pose more serious and immediate threats to health, even in small quantities.

The CAA was reauthorized in 1977 and, after more than a decade of political struggle, again in 1990.[4] Both versions largely reaffirmed the earlier strategy. The 1977 act incorporated the emissions offset policy that EPA already had adopted administratively (discussed in chap. 6).[5] The 1990 version (which I will call the CAAA, for Clean Air Act Amendments) was more important.

Several aspects of the CAAA stand out. First, it revised the provisions for attaining and maintaining the NAAQS for ozone, carbon monoxide, and fine particulates. The most important provision dealt with ozone, for which (based on 1987–1989 data) nearly one hundred areas in the country had not attained the NAAQS. These nonattainment areas were placed in five groups, according to the severity of their ozone pollution; these ranged from "marginal" (39 areas) and "moderate" (32 areas) to "serious" (16 areas), "severe" (8 areas), and "extreme" areas (Los Angeles, which was literally in a class by itself). The severe areas were given more time to attain the NAAQS but had to meet more stringent and a greater number of control requirements. State Implementation Plans had to specify how the ozone NAAQS would be attained. Plans for the dirtier areas had to include tighter controls, cover more sources, use economic incentive programs, and require more controls on mobile sources and on transportation than did the plans for the other areas. Areas that failed to attain the NAAQS for carbon monoxide and fine particulates were classified as either "moderate" or "serious." Areas classed as "serious" had to adopt more stringent controls than the "moderate" ones: enhanced auto inspection and maintenance, driving restrictions, and strict industrial controls.

Second, the law expanded the limits on emissions from motor vehicles and set standards for a newer generation of clean fuels. Areas with high ozone levels had to begin using reformulated gasoline and expand the use of fleet vehicles (such as taxis and delivery vans) that could run on these cleaner fuels. Congress set tighter limits on tailpipe emissions that would lead to lower levels of hydrocarbons, carbon monoxide, nitrogen oxide, and particles in the ambient air. EPA must set even more stringent limits on vehicle emissions by the year 2000, if that appears necessary for achieving the NAAQS.

The 1990 law also adopted strong new provisions for air toxics. EPA's inability to set more than a handful of standards for toxic air pollutants under the earlier law led Congress to list pollutants and insert deadlines. The law directs EPA to set Maximum Available Control Technology (MACT) standards for a list of 189 toxic air pollutants for categories of sources (such as petroleum refineries) emitting those pollutants. MACT standards have to be set to reach the maximum emission reductions that EPA decides are achievable, taking cost and other factors into account, and must be stricter for new than for old sources. In addition, if EPA determines that significant residual risks remain after these technology-based controls are in place, it must require even stricter controls that protect health with an ample margin of safety.

The fourth major aspect of the CAAA dealt with acid rain control. Large industrial sources—mostly utilities in the Midwest—emit sulfur oxide and nitrogen oxide from tall smokestacks. These pollutants may be transported long distances and chemically transformed, then fall as acid rain over New England and Canada. This title of the CAAA set a goal: to reduce annual emissions of sulfur oxide to ten million tons and nitrogen oxide to two million tons below their 1980 levels. Most of these reductions would be achieved through an innovative allowance trading system (discussed in chap. 6).

WATER QUALITY

Themes from the CAA of 1970 were played again in the Federal Water Pollution Control Act (FWPCA) of 1972. Earlier water laws (enacted in 1948, 1956, and 1965) had limited the federal role to guidance, research, investigation, and grants. The 1972 law changed this. Congress directed EPA to set uniform, national limits on effluents for all major sources of discharges into water. These limits, known under the law as "effluent guidelines," are similar to the NSPS in air. They were designed to force industry to adopt the latest available technology, with a few caveats. Though states had a role, the water act, like the CAA, set up a centralized program with federal oversight of state activities, at least for point sources of pollution (those with an identifiable discharge point).[6]

Yet the air and water acts differ. The goals in the FWPCA were (1) to eliminate all discharges into navigable waters of the United States by 1985, and (2) to make all water "fishable and swimmable" by the middle of 1983. Both were ambitious goals that still have not been met. The FWPCA expanded an existing federal program of grants to municipalities for the construction of sewage treatment plants. Congress authorized the federal government to pay 75 percent of the cost of local treatment works. There is nothing comparable in the air program, where nearly all costs are borne by the private sector.

The air and water acts also differ in their use of ambient standards to define goals and set limits on dischargers. Under the water program, there is nothing equivalent to the health-based, nationally uniform NAAQS. Earlier water laws had made states set water quality standards for specific bodies of water (lakes or segments of rivers), then draft plans for meeting them. Standards varied, depending on the state's decisions about the "designated use" of the water. State permit officials were to assign limits to dischargers as necessary to keep levels of contaminants at or below the standard. The FWPCA continued this program, with more federal oversight. States must review their standards every three years and submit any changes to EPA for approval. They may set stricter standards based on higher (cleaner) designated uses, but any relaxations of the standards are carefully reviewed and may be rejected by EPA.

The FWPCA also added a system of technology-based effluent guidelines to the program, to apply to industrial dischargers and municipal sewage treatment plants. These set numerical limits based on EPA's judgments about "currently available" (by 1977) or "best available" (by 1983) control technologies that dischargers could be expected to meet. The only qualification was that the limits and technologies they are based on must be "economically achievable." Like the NSPS, these effluent guidelines apply to all new sources in defined categories of industrial dischargers, even if the receiving water meets the local quality standard. If by meeting the guideline a state still does not achieve the standard, EPA or the state can require still stricter controls. So although the water act relies on ambient standards, they vary by location and the designated use prescribed for a water body. Like the air act, a failure to meet a standard can trigger stricter controls.

Congress has reauthorized the water act twice since 1972. The 1977 revisions postponed some deadlines and gave more weight to the control of toxic pollutants. The Water Quality Act of 1987 (passed over President Reagan's veto) again put off some of the deadlines and expanded federal programs for nonpoint sources, like discharges from urban storm sewers. The 1987 act changed the approach for financing wastewater treatment, from a system based on grants to local governments to a state revolving loan fund. The law also expanded water quality programs for lakes and estuaries, including special programs for the Great Lakes and other areas.

HAZARDOUS AND SOLID WASTES

The first significant national law to deal with hazardous and solid wastes came out of Congress in 1976, after the air and water laws were passed. The legal framework grew rapidly after concern about the health effects of hazardous wastes rose quickly on the national policy agenda at the end of the 1970s. Hazardous waste then became one of the big environmental issues of the 1980s.[7]

The key event was passage of the Resource Conservation and Recovery Act (RCRA) in 1976, superseding the Solid Waste Disposal Act of 1965. The RCRA set a framework for regulating the generation, transportation, treatment, and disposal of hazardous wastes. The goal was to create a cradle to grave system for managing wastes. Under the RCRA, EPA issued regulations that defined hazardous wastes; tracked wastes from generation to disposal; and set technical standards and rules for issuing permits to facilities that stored, treated, or disposed of wastes.

In the 1976 version, EPA had discretion in deciding on specific provisions. This changed in 1984. Congress reauthorized the RCRA with the Hazardous and Solid Waste Amendments (HSWA). EPA's slow pace in issuing the RCRA rules, the Reagan administration's disregard for environmental problems, and public concern about hazardous waste led Congress to pass a very prescriptive law. It directed EPA to regulate new classes of facilities, including generators of small quantities of hazardous wastes and underground petroleum and chemical storage tanks. To reduce the chance of groundwater contamination, the law set strict limits on the disposal of wastes on land. The HSWA dealt with facilities that close and leave waste behind by establishing liability, insurance, and "corrective action" provisions to cover cleanups.

Even more significant is how Congress wrote the HSWA to force action and severely limit EPA's discretion. Many provisions look more like regulations than legislation. Because the legislative deadlines under previous laws had been missed, Congress wrote "hammers" into the HSWA. Under the hammers, if EPA failed to issue rules by a given date, the statutory requirements automatically fell into place. The most important such provision in the HSWA is a presumptive ban on land disposal of hazardous wastes.[8] In the HSWA, Congress gave little attention to the costs and economic impacts of the program it was directing EPA to implement. Because the RCRA does not explicitly allow EPA to consider costs in setting standards, the HSWA rules are among the most expensive that EPA has issued.

The RCRA/HSWA regulatory scheme addressed risks from active waste sites. Its aim was to prevent future Love Canals, not clean up the existing ones. To address problems from inactive waste sites, Congress passed the Comprehensive Environmental Response, Compensation, and Liability Act (CERCLA, also known as Superfund) in 1980.[9] This was the first environmental law designed to remedy problems that were created in the past rather than reduce pollution as it is created, and it presented a new set of issues and challenges. It gives EPA authority to identify the parties responsible for waste sites and to force them to clean up. Costs are covered either by the responsible parties or by a trust fund. The CERCLA directs EPA to identify sites, rank them according to the hazards they present, and maintain a National Priority List (NPL) for cleanups. Congress amended Superfund in 1986 and added community-right-to-know provisions to the law. These provisions require firms to maintain and make available data about harmful chemicals that are used or stored on a site and to report the annual emission of such chemicals. This Toxics Release Inventory (TRI) has become a major source of data on industrial emissions.[10]

PESTICIDES AND TOXICS

First passed in 1947, the Federal Insecticides, Fungicides, and Rodenticides Act (FIFRA) authorized a national registration program for pesticides. This law stressed the effectiveness of a product as the basis for a decision on registration, but later versions focused more on a chemical's effects on health and the environment. The core of EPA's regulatory authority is that only EPA-registered products may enter commerce. The current standard for registration is that, under normal use, a chemical will not cause "unreasonable adverse effects on the environment." In making this decision, EPA must consider "the economic, social, and environmental costs and benefits of the use of any pesticide." The law directs EPA to balance risks and benefits; higher economic benefits may offset higher risks.

Like many environmental laws, the FIFRA sets higher standards for new than for existing products. For new products, the burden of proof is on registrants to show they are safe. For existing products, the burden is on EPA to show unreasonable risk before canceling or suspending registration. These parts of the law were defined in the 1972 revisions and amended in 1988. The 1988 changes directed EPA to reregister six hundred active ingredients in pesticides (those with the older registrations) within nine years.

One aspect of pesticides is governed by the Federal Food, Drug, and Cosmetic Act (FFDCA). Here EPA sets tolerances, or maximum allowable concentrations, for pesticide residues in food. The law defines one standard for EPA to use in setting tolerances on raw agricultural commodities and another, stricter standard for setting tolerances on processed foods. The differences do not reflect a difference in the risks of the two kinds of products. They reflect political and legal issues in the passage of the law. This difference in how raw commodities and processed foods are treated is one of the obvious discrepancies in the legal framework.[11]

Passed in 1976, the Toxic Substances Control Act (TSCA) was designed to cover a range of chemicals currently in production and use that could present health or environmental risks and were not covered by other laws. It gives EPA broad authority to collect information on chemical substances and mixtures, to require industry to test chemicals for harmful effects, and to regulate the production or use of any chemicals that pose "an unreasonable risk of injury to health or the environment." EPA maintains an inventory of existing chemicals (now including some 60,000), for which it can require testing or take regulatory action if there is evidence of an unreasonable risk. In deciding whether to regulate, EPA must balance the probability and significance of harm from a substance against its benefits and the availability and risks of substitutes. Like the FIFRA, the TSCA directs EPA to balance benefits against risks in regulatory decisions.

THE POLLUTION PREVENTION ACT

The Pollution Prevention Act of 1990 was a departure from the established style of environmental regulation. It reflected the recognition that this country had relied too much on end-of-pipe controls and too little on a strategy of preventing pollution at the source. The law defined as national policy that (1) when it was feasible, pollution should be prevented or reduced at the source; (2) when prevention was not feasible, pollution should be recycled in an environmentally safe manner; (3) when prevention or recycling was not feasible, pollution should be treated in environmentally safe ways; and (4) disposal or releases of pollution into the environment should be used as a last resort—when prevention, recycling, or treatment is infeasible. Prevention is defined as "source reduction"—to include anything that increases efficiency in the use of raw materials, energy, water, and other resources, or protects resources through conservation.

What is most notable about the Pollution Prevention Act is that it is not limited to one environmental medium (i.e., air, water, or waste) or program. Indeed, within EPA and on Capitol Hill, there has been a careful effort not to limit the pollution prevention concept to traditional program boundaries. Unlike the earlier environmental laws, the Pollution Prevention Act promoted a more integrated approach to environmental regulation.

OBSERVATIONS ON THE STATUTORY SETTING

All of this constitutes a formidable and complex legal setting for policy making. Just gaining an acquaintance with the range and diversity of environmental laws could occupy a big chunk of one's professional experience in the environmental field. Here are three observations about the statutory framework before we move on.

(1) *Changes in environmental laws over time have been incremental.* Incremental policy systems change slowly. Their basic structure varies only slightly over time; when change does occur, it is at the margins, and it builds on experience.

From the beginning, the basic pattern in the statutes has been to rely on command-and-control regulation, usually with technology-based standards. Despite all the attention given to the various provisions for economic incentives in the 1990 CAAA, the law as a whole is strongly oriented toward traditional, technology-based regulation. Despite growing concern about the fragmentation in legal authorities, Congress has retained the old divisions among the air, water, and other programs. With the exception of the Pollution Prevention Act, the incremental pattern has continued.

The CAAA is a good example. Those amendments, Arnold Reitze has written, did what environmental laws in the United States have always done: regulate "what comes from smokestacks without dealing effectively with the population and consumption issues that stress our atmosphere and threaten the equilibrium of our biosphere."[12] The pattern is to require ever more stringent regulation without dealing with the roots of problems, which can only be addressed through more comprehensive land use and energy policies. It is classic incremental policy making.

(2) *Environmental laws have become more and more prescriptive.* There was a time when administrative and legal theorists worried most about the vague delegations of authority given to agencies in their enabling laws.[13] The pattern in the environmental laws passed over the past two decades stood this old worry on its head. The issue today is not vague delegations but very prescriptive laws that leave little room for agency discretion. The HSWA of 1984 set a trend, with its many deadlines, hammers, and specific regulatory requirements. The 1987 water act continued the trend, and the 1990 CAAA was probably the most prescriptive yet.

Why the trend toward prescription and away from delegation? The major reason Congress limited EPA's discretion and pushed it toward more stringent and expensive regulation over the past ten years was the environmental performance of the Reagan and Bush administrations and the divisions between Congress and the White House that persisted through the decade. Put simply, Congress limited EPA's discretion because it did not trust the White House to implement the law.

This concern is expressed by Rep. Henry Waxman, a chief proponent of many of the stringent provisions in the 1990 CAAA. These statutory deadlines and mandatory provisions, he has written, "are routinely provided to assure that regulatory actions are taken in a timely fashion."[14] For example, the law lists 189 air toxics to be regulated, with timetables for action on each, giving EPA limited discretion in setting standards. For mobile sources, the law eliminates EPA's discretion almost completely by setting more than ninety specific emission limits for vehicles. This reflects a concern that "without detailed directives, industry intervention might frustrate efforts to put pollution control steps in place."[15] If the executive branch will not carry out the laws, then laws had to be written by Congress to be almost self-executing.

The irony is that prescriptive laws with strict deadlines and stringent regulatory provisions probably will increase the overall cost of environmental regulation to American society. This is precisely the effect the Reagan and Bush administrations wanted to avoid.

(3) *The statutory framework reflects an ambivalence, if not hostility, toward economics.* The tensions between economic and environmental values have been a central theme in U.S. policy. Anyone looking for a consistent treatment of economic analysis across the different laws will search in vain. Policy makers have resolved the conflicts between the nation's environmental and economic goals by muddling through, even within the same law.

The RCRA directs EPA to set standards that will "protect human health and the environment," without suggesting whether the costs of regulation are relevant. The CAA requires NAAQS that are strict enough to protect the most sensitive groups in the population, whatever the cost. In setting the NSPS or effluent guidelines, EPA must consider control costs and effects on sources but not the benefits of reducing emissions. The TSCA and the FIFRA both use an economically rational policy model by directing EPA to balance the costs and benefits of controls. But the former has been used very little to regulate existing chemicals, in part because of the difficulty of making a cost-benefit case that can withstand legal and political challenge. And the latter affects a relatively narrow segment of the economy; its effects on the costs and benefits of environmental programs overall are marginal.

If we look beyond the regulatory statutes themselves, the ambiguities and conflicts are even more apparent. The Regulatory Flexibility Act, passed in 1980, directs agencies to minimize the adverse economic impacts of rules on small firms and communities, "to the extent permitted by law." Because this law must give way to the other regulatory laws, however, it has had a marginal effect on policy. Expressing a very different sentiment, the Endangered Species Act prohibits an agency from taking an action that would threaten an endangered species, without regard to economic effects. Yet, responding to criticism that the act went too far in protecting endangered species, Congress amended it in 1978 to allow a Cabinet-level Endangered Species Committee to exempt specific projects from its limitations.[16]

The statutory framework for national environmental policy is both complex and internally inconsistent. Although it has been effective in allowing EPA to build a strong regulatory base for reducing many of the harmful effects of industrial activity on American society—in itself no mean achievement—this framework has also led to the creation of an elaborate and often expensive body of regulation. We turn now to the institution that has had the job of interpreting, implementing, and sometimes shaping this complex statutory framework.

EPA AS THE FOCUS FOR POLICY MAKING

Throughout American history, Congress has chosen to carry out its will by giving authority to administrative agencies. In the early days of administrative power, the reasoning was simple: Congress was too busy and too much an institution of generalists to have the technical knowledge needed to regulate on complex issues; agencies were seen as the logical way to extend the capacities of Congress and carry out its wishes in a nonpolitical and expert manner. And so, for example, Congress created the Interstate Commerce Commission to regulate railroads, the Food and Drug Administration to monitor food additives and drugs, and the Federal Trade Commission to administer antitrust laws.[17]

Experience has shown, though, that however technically expert agencies may be, they can hardly remain apolitical and impartial. They necessarily are political animals: they need to make choices among competing values that often are in conflict; they have to cultivate and respond to the demands of political constituencies; their decisions provide grist for the mill of election campaigns. We now are more realistic about the limits in agencies' technical knowledge and the difficult value choices they must make. Deciding whether to ban chemicals or to allow construction of major water projects that destroy ecologically valuable wetlands requires more than just technical expertise. It requires agencies to make choices among competing values and to defend their choices on legal, political, ethical, and scientific grounds. EPA and the framework of laws surveyed here reflect the recognition that environmental problems require a national response. Like most agencies, EPA was born of and operates within an intensely political reality.

My survey takes in three topics: EPA's formation and history; its organization and budget; and institutional perspectives that affect policy making. I then look at EPA's rule making process.

EPA'S ORIGINS AND HISTORY

EPA's beginnings under President Richard Nixon came in the form of an executive branch reorganization in December 1970.[18] Three aspects of its birth are worth noting. First, the formation of a national agency was one of many governmental responses to concern about the environment that was sweeping the United States and the world at the start of the 1970s. In 1969, Congress passed the National Environmental Policy Act, a major innovation in its own right. The nation marked the first Earth Day only four months after EPA was created. Major new environmental laws were enacted in the next few years. The 1970s were truly an environmental decade, when the institutional mold for national policy was cast.

Second, the Nixon administration made EPA a separate pollution control agency, not part (with the Department of the Interior) of a larger department of natural resources, as many had advocated. This was more than a cosmetic issue; environmental policy would look very different today if pollution programs had been placed in a large department that stressed developing more than conserving resources. Conservation-oriented agencies within the Department of the Interior, such as the Fish and Wildlife Service, struggle to push their environmental agendas. EPA has its own internal conflicts, but they are fewer than they would have been under a department that was responsible for managing both resource development *and* environmental protection.

Third, EPA came about through an executive reorganization, rather than through passage of an organic environmental statute that defined its overall authorities. The result was to bring several existing agencies and laws together into an environmental holding company, not to create an integrated agency with a clear and internally consistent set of programs. EPA was patched together from the Departments of Health, Education, and Welfare (air quality, solid waste, drinking water), Interior (water quality, pesticides research), and Agriculture (pesticides regulation), and other agencies (the Food and Drug Administration and the Atomic Energy Commission). EPA was "not a single organism with a single will but a series of different organisms with different wills."[19]

Good discussions of EPA at different stages in its history exist, but a general history of the agency has yet to be written.[20] For our purposes, I review the major stages in its history.

The Early Years: Establishing a Presence Under its first administrator, William Ruckelshaus, EPA set out to define an aggressive enforcement profile. The environmental movement was in full bloom, and EPA's leadership saw a need to quickly establish itself and show results against the large industrial sources of air and water pollution. The major statutes concerning air and water pollution emerged from Congress during this period. Both gave Ruckelshaus the clout to play a strong role, despite the caution of the Nixon administration. The goal was for EPA to "hit the ground running" by issuing rules and enforcing them as visibly as possible. Ruckelshaus made EPA a high-profile agency, a profile that did not always sell well with inflation and energy shortages in the 1970s. Yet EPA was firmly set on the institutional landscape; it developed a culture of environmental activism that persists to this day, despite the political changes over the years.

One crucial set of decisions that Ruckelshaus faced concerned EPA's organization. Because the agency had been pieced together from several agencies, many argued for a reorganization that would structure it more along functional (permitting, monitoring) than program (waste, water, air) lines. Ruckelshaus decided to postpone any change, however, thinking that it would be too disruptive at this early stage in the agency's history.

Years of Consolidation and Reform A period of consolidation and growth in technical competence in the mid-1970s was followed by a time of expansion in the legal authorities and a movement toward regulatory reform. The passage of the TSCA and the RCRA in 1976 brought new issues onto EPA's agenda and greatly expanded its portfolio. Russell Train moved from chairing the Council on Environmental Quality to become administrator in 1974. Like Ruckelshaus, Train was a moderate Republican who fought to maintain EPA's independence and its commitment to environmental values, despite a skeptical White House. Train offset the influence of the White House by cultivating close ties with supporters in Congress, especially Sen. Edmund Muskie's environmental pollution subcommittee. He also decided not to make fundamental organizational changes; the "untidy mix of functional and program offices was retained."[21]

President Jimmy Carter named Douglas Costle to head the agency in 1977. The energy crisis, high inflation, and the troubled economy of the 1970s brought environmental policies under scrutiny. As a result, many proposals for regulatory reform emerged during these years. Although actual deregulation in this period was limited to the economic regulatory functions of agencies like the Interstate Commerce Commission (ICC) and the Civil Aeronautics Board (CAB), environmental regulation was the subject of debate as well.[22]

The Costle years combined expanded regulation (especially in hazardous wastes) with a movement to reform regulation and limit its economic effects. Carter issued an executive order requiring more economic and cost-benefit analysis from regulators, created the Regulatory Analysis Review Group in the White House to analyze the economic effects of regulatory actions, and increased the oversight authority of the Office of Management and Budget. Carter also supported two laws for limiting regulatory burdens: the Regulatory Flexibility Act, which directed agencies to minimize economic impacts on small businesses and governments, and the Paperwork Reduction Act, which required agencies to analyze and justify reporting and record-keeping burdens that they imposed on the private sector.

It was also under Costle that EPA began to cast itself more as a health than an environmental agency, especially in its efforts to identify and reduce the risks from cancer. This public health, anticancer emphasis was seen at the time as a necessary political adaptation to the antiregulatory pressures of the day. By casting itself as a defender of the public's health against the scourge of cancer, EPA was able to cultivate more support for its programs. It also began to stress risks from toxic pollutants that posed the more serious, direct threats to health—like asbestos or dioxin—over those from the conventional pollutants that had been its focus in the 1970s. It was not until the late 1980s that ecological issues again began to compete with health issues on EPA's agenda.

Years of Retrenchment and Renewal Until 1981, EPA had seen years of steady growth in resources and statutory authority. This ended abruptly in 1981, when President Ronald Reagan came to Washington with what he thought was a strong mandate to reduce the costs and effects of government regulation. Because of the scope and growing costs of its regulations, EPA was a special target. Administrator Anne Gorsuch set out to reduce EPA's budget and programs, delegate more authority to state and local governments (the New Federalism), and cut back the scope and stringency of regulation.[23]

The atmosphere at EPA during the early 1980s can fairly be described as dismal. Never before in recent American history had an administration come to office professing outright hostility to regulation, with environmental rules as a prime target. Reagan appointed Gorsuch, a little-known Colorado state legislator with almost no environmental or management experience, to run the agency. The second rung of management—assistant administrators (AAs)—was filled with people who had weak management and policy skills and a suspicion of environmental programs. The new leadership tried to insulate itself from the career staff at EPA; it compiled lists of "suspect" officials (those seen to be too environmental in their sympathies), rejected staff advice on most issues, and catered to industry influences and pressures, often in violation of administrative procedures. As budgets shrank and reliance on staff expertise fell, the morale of agency staff declined.

The early 1980s illustrate that policy change can only go so far when political support is lacking. The most direct way to recast national policy would have been to change the laws. After all, environmental statutes defined the basis for EPA's regulatory authority and for oversight by Congress and the courts. By changing the laws, the administration could have made the radical changes it thought it had an electoral mandate to accomplish. Yet the Reagan team failed in efforts at legislative change. Public support for the environment was far stronger than they had anticipated, and Congress showed that it would not give up the gains of the previous decade. With its legislative agenda frustrated, the Reagan administration sought change through an administrative and budgetary strategy.

The Reagan administration misread the level of political support for environmental protection, and the Gorsuch team made fatal errors in managing EPA and its relations with Congress. By March 1983, nearly all of EPA's political leadership had resigned. In pursuit of its "mistaken mandate" to dismantle environmental regulation, the Reagan team had disqualified themselves and the president from claims to leadership on environmental policy.[24]

Ruckelshaus returned in the spring of 1983 to rebuild the agency and its credibility. He worked to restore EPA's budget, morale, and enforcement record and improve strained relations with Congress.[25] Aside from his political restoration of EPA, perhaps Ruckelshaus's major legacy was to promote the debate over risk and its role in policy. A consistent theme of his leadership was that not all threats to health or the environment can be controlled by government action. There are, as one of his aides wrote, "a near infinity of potential targets" calling for attention.[26] Society's resources, he argued, should be directed toward the problems that threaten the most harm and present the best opportunities for reducing risks. It was under Ruckelshaus and his deputy, Alvin Alm, that EPA began its *Unfinished Business* report, which compared and ranked sources of health and ecological risk on EPA's agenda.[27]

Ruckelshaus and his successor, Lee Thomas (who took over in 1985), worked in an administration whose support for EPA programs was lukewarm, at best, and an OMB still bent on limiting costs. At the same time, Congress was pulling in another direction—toward more stringent and prescriptive laws in the reauthorized Superfund, RCRA, water quality, and drinking water statutes. Despite greater congressional confidence in EPA under Ruckelshaus and Thomas, congressional distrust of the Reagan administration led to close oversight and increased executive–legislative conflict.

The Reilly Years President George Bush's appointment of William Reilly as EPA administrator in 1989 moved EPA and national policy in several subtle but important new directions. International issues became a major theme, in response to the emergence of such problems as depletion of the ozone layer, global warming, long-range air transport, waste exports, regional oceans and water quality issues, and more economic interdependence among nations. Another theme was the emphasis on ecological as well as health risks, with terms like "biodiversity," "wetlands," and "habitat protection" getting more play. The Reilly team also continued the process that had begun under Ruckelshaus and Thomas, emphasizing risk in setting priorities and making decisions.

This period was also marked by a growing visibility for EPA within the administration and probably the closest ties between an administrator and president in its history, as well as by proposals to elevate EPA to Cabinet status. Yet the Cabinet status bill never passed, due to squabbles about minor provisions attached to it. And by the end of President Bush's term, Reilly's relations with the White House were anything but smooth. The president had decided to stress economics over the environment in his reelection bid. EPA and Reilly came up short. The president pulled away from his support for a "no net loss" wetlands policy, niggled at regulations implementing the CAAA, and undercut efforts to allow the United States to play a leadership role at the Rio de Janeiro Earth Summit. In March 1992, *The Wall Street Journal* noted, "White House officials second-guess even the tiniest details of EPA proposals, leading to drawn-out battles over how wet a wetland must be and what kind of boilers should be in power plants."[28]

The Clinton Administration and Carol Browner In his first round of appointments during the 1992–1993 transition, then President-elect Bill Clinton selected Carol M. Browner, head of the Florida Department of Environmental Regulation, as the EPA administrator. Earlier in her career, Browner had worked as an aide to Sen. Al Gore, and her selection was seen as a sign of Vice President Gore's likely influence on environmental policy. In her early statements of priorities, Administrator Browner stressed as key themes pollution prevention, ecosystem protection, environmental justice (attention to the risks of environmental problems for minorities), innovation in environmental technologies, and partnerships (improved working relationships) with states.[29] The issue of environmental justice became particularly important in her first year in office, as many grass-roots groups and other critics pushed EPA to reexamine national policies regarding the siting of waste facilities, the cleanup of Superfund sites in minority and low-income communities, exposures to lead and other inner-city health problems, and the effects of pesticides on migrant workers.

A major symbolic position of the Clinton administration was its support for a bill that would make EPA a Cabinet agency, as the Department of Environmental Protection. Although there was broad support for the Cabinet-status bill in principle in Congress, many legislators wanted to use it as a vehicle for passing amendments that would require EPA to conduct a cost-benefit and risk analysis for all regulatory decisions, a requirement that the Clinton administration opposed. The Senate passed a bill in May 1993 that would have elevated EPA to Cabinet status, but this version included the amendment requiring across-the-board cost-benefit and risk analysis, sponsored by Sen. J. Bennett Johnston of Louisiana. The administration was unable to get the House to consider the Cabinet bill without this provision, and there was no further progress.[30] As of mid-1994, the bill was stalled in Congress, although it still was possible that EPA would become the Department of Environmental Protection in 1994 or 1995.

As part of the White House staff, President Clinton created the Office of Environmental Policy (also headed by a former Senate aide of Al Gore) to coordinate policy across agencies. This group has taken the lead in the administration on such issues as global warming, wetlands, pesticides in foods, and environmental aspects of the North American Free Trade Agreement (NAFTA) with Mexico.

ORGANIZATION AND BUDGET

Like most large agencies, EPA combines substantive, functional, and geographic principles of organization. The substantive divisions of labor fall mainly along program lines. Responsibility for the national programs (such as air, water, and waste) lies with the AAs for Air and Radiation; Water; Solid Waste and Emergency Response; and Prevention, Pesticides, and Toxic Substances. Offices based on functions are led by the AAs for Policy, Planning, and Evaluation; Research and Development; Enforcement and Compliance Assurance; International Activities; Administration and Resources Management; and the General Counsel. Other staff offices handle legislative and state/local affairs, public relations, and special issues. Like the administrator and deputy administrator, the AAs are appointed by the president and confirmed by the Senate. The senior career civil service person in each office is usually a deputy assistant administrator; the operating offices below them are headed by office and division directors (who are career staff). Regional administrators (who are political appointees) head EPA's ten regional offices. Figure 1 presents a picture of EPA's organization as it looked in 1994.

EPA's staff in 1994 numbered some 18,500. It is a highly professional group; two-thirds have college degrees, and one-third have graduate degrees. The largest single discipline represented is engineering, followed by law and business; agricultural, health, and biological sciences; physical and environmental sciences; and the social sciences. Agency staffing declined from about 13,000 in 1981, at the start of the Reagan years, to about 11,000 in 1983, before rising steadily through the decade to its current levels. Most of the growth was in Superfund. For the 1994 fiscal year, EPA's total budget was $6.6 billion. In current dollars, this is slightly above the funding level for 1981; so in real terms,

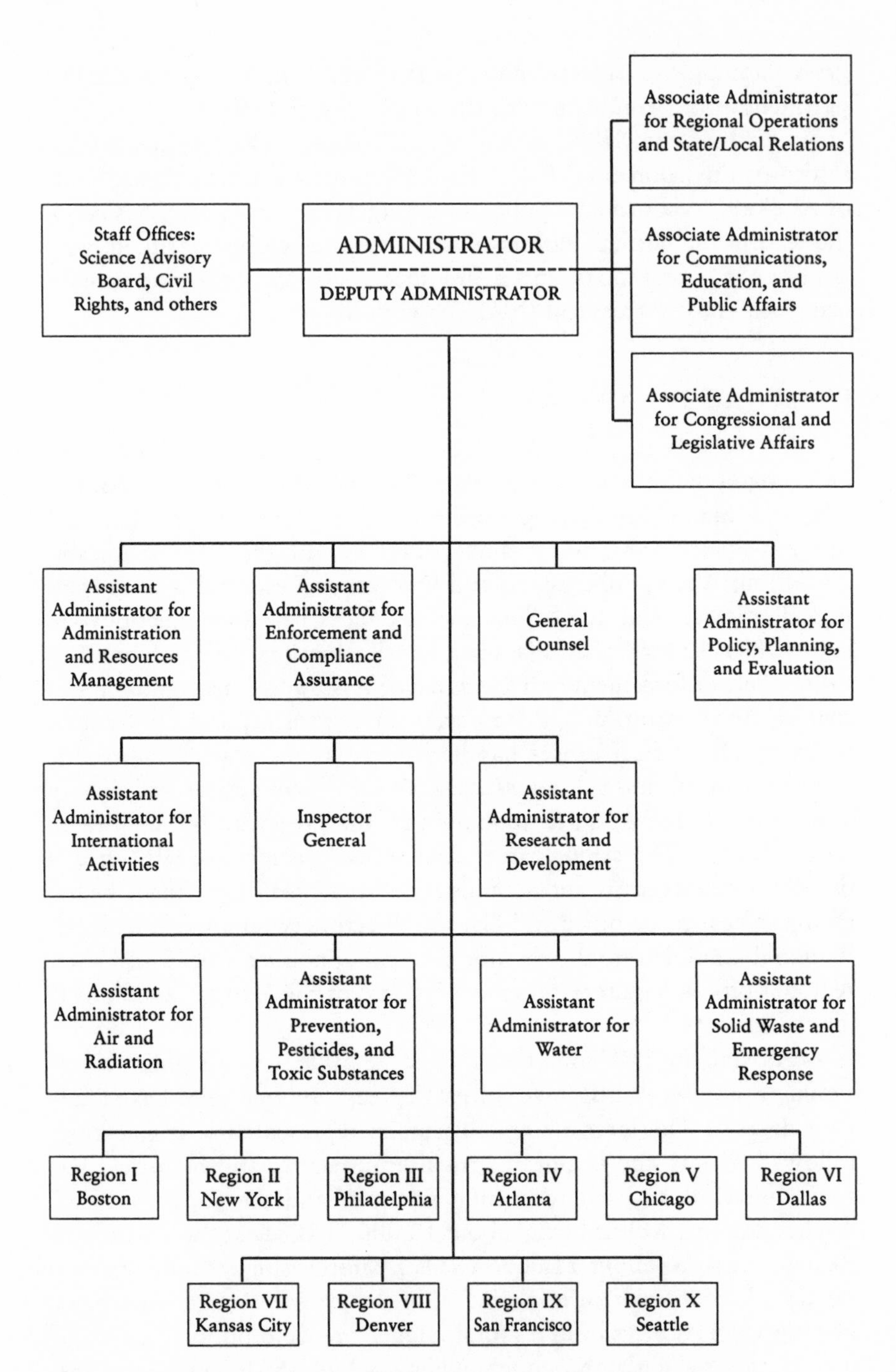

Figure 1. Organization of the U.S. Environmental Protection Agency.

EPA lost ground in funding and staff over the decade, despite an expansion of its authorities. The operating budget (what was available directly to EPA) in 1994 was about $2.7 billion. Other big pieces of the budget were Superfund ($1.5 billion) and water construction projects, mainly the Clean Water State Revolving Fund ($2.4 billion). Within the operating budget, some $600 million goes to state/local governments for special purpose programs and activities, such as the drinking water program, the wetlands program, and pesticides enforcement.[31]

INSTITUTIONAL PERSPECTIVES THAT AFFECT POLICY

Information about organization, staffing, and budgets tells only part of the story. Agencies are not well-oiled machines but collections of interests, roles, and specialists, each with its own vision of the agency's mission and its own core values. By looking at different perspectives, we gain a better understanding of how policy is made and why. In a book about EPA's early years, Alfred Marcus contrasts program, policy, and research perspectives.[32] More recently, William West distinguishes the perspectives of program staff, executives, policy analysts, and the general counsel's office.[33] I will distinguish five institutional perspectives within EPA: program, policy, legal, research, and regional.

The national program offices formulate and implement policy in the air, water, waste, and other areas. They are oriented strongly toward the enabling statutes that define their mission, and they tend to have strong ties with outside groups that share their interests and have a stake in their programs: the legislative oversight committees, state and local agencies, and environmental groups or trade associations. Within the drinking water program, for example, there are strong ties with the American Water Works Association (AWWA) and the House and Senate subcommittees that oversee the program. There also is vertical integration among levels of government—federal, state, and local—that regulate or manage drinking water supply systems. There are differences in professional outlook as well; program staff are more likely to have technical training in engineering, toxicology, and science than are, for example, policy staff.

Next policy, legal, and research perspectives can be defined. Staff from the Office of Policy, Planning, and Evaluation (OPPE) take an analytical approach to the agency's work. They emphasize the analysis of costs and benefits, cost-effectiveness, and economic impacts; cross-media planning; and the use of relative risk to set priorities and allocate the social costs of programs. Their training is usually in economics, public policy, or the social sciences, although many also have technical skills. A fact of life for the policy office is that it lacks a natural constituency outside of EPA. Although its views are often more allied with those of the Office of Management and Budget than are the program's, they differ strongly from OMB's in many cases. Environmentalists are skeptical of the policy staff's concerns about costs; oversight committees resent the lack of commitment to their program. The main client for the policy office has usually been the agency's policy leadership. OPPE's role waxes and wanes, depending on its ties with and value to the administrator and the deputy.

As one would expect, the professional ranks of the general counsel's office are made up of lawyers. As the courts have become more involved in regulatory policy, the influence of legal staff has grown. They stress interpretations of enabling laws and the legal defensibility of decisions, especially regulatory ones. They are strongly influenced by the statutory text and the legislative record that supports it, trends in court review, and procedural concerns. The role of legal staff is pervasive; they take part in nearly all rule makings and have a de facto veto on any decisions through their interpretation of statutory authorities.

Another perspective is that of the Office of Research and Development (ORD), which places a high value on the quality of the agency's scientific work and the consistent application of scientific methods and analysis across programs. Made up of staff who are trained mostly in the physical and chemical sciences or engineering, this office directs research on assessing health and ecological risks, technologies to control pollution, ways to measure and monitor pollutants, and others. The research is done in EPA labs around the country or by contractors and universities. A criticism of ORD has been that its research is not adequately linked to the priorities and near-term needs of the policy makers. ORD does affect decisions, however, by doing risk assessments on which policy decisions are based (e.g., health assessments for the NAAQS) and assessing plans (e.g., monitoring plans for measuring contamination of groundwater by pesticides).

Yet another perspective is that of the regional offices. Regions play several important roles: as links between state-level environmental agencies and Washington; as adapters of national policies to regional conditions; as voices for practicality and common sense; as agents of the national programs and overseer of state compliance with their requirements. They are also microcosms of EPA headquarters; their offices are divided according to environmental media, and their legal and enforcement staffs correspond to the staffs in Washington.

National policy matters most to regions when it affects the allocation of responsibilities among levels of government, the distribution of agency resources, or the degree of discretion that regional and state officials can exercise. Attention to problems varies across regions, as does regional interest in national issues: lead is big in Boston, with its old housing stock; wetlands are an issue in the Southeast; radon emerged as a priority in Region III (Philadelphia) because of the Reading Prong area in the eastern part of the state, where radon levels are unusually high.

These differences in roles and perspectives are a source of institutional tension. Often, program and policy offices disagree about the costs of a policy compared to its benefits, the program pushing for a stringent standard and the policy office favoring a less stringent but less costly (or more cost-effective) one. Program staff may favor centralized national programs, while the regions argue for regional flexibility. Legal staff may oppose an option favored by program and policy staffs because it does not conform to their interpretation of the statute or the legislative record. Program, policy, and legal staffs may agree on a decision to treat a risk as minor, but the research office may be concerned that the health assessment does not *prove* the absence of risk.

Are these institutional tensions desirable? A short answer is that they are, and to some degree they are deliberately built into the structure of EPA and most large policy making agencies. Whether consciously or not, most agencies incorporate a degree of multiple advocacy in their designs. In making decisions as large and complex as setting national health standards for ambient air, technical rules for designing and maintaining waste landfills, or permitting rules for sources discharging into storm water sewers, most decision makers want multiple sources of advice and analysis. This reduces the chances that decisions will be based on weak data or invalid assumptions. It also brings into an agency the multiple values that environmental policies force us as a society to confront—of economic versus environmental goals, the need for good science against the need to make decisions, the balance of regional flexibility against national uniformity, or growth and development versus the preservation of resources. In this sense, EPA's internal tensions reflect the checks and balances present in the American political system.

EPA'S ROLE

EPA operates in a complex and controversial institutional setting. It also works under institutional handicaps, among them the lack of a single organic statute and its origins as a patchwork of organizations and programs that were shifted from other agencies.

Even within EPA, debate has been growing about what its role should be. Should it continue mainly as a national regulator, with standard setting and enforcement its most visible activities? Or should it work toward a new conception that recognizes the growing complexity of environmental issues and the limits of a purely regulatory strategy? This new conception pictures EPA as less a regulator and more a source of education, technical advice, technological innovation, and international leadership. The old pattern of conflict with industry would be replaced by a new relationship based on cooperation and joint problem solving. Instead of deciding what technology-forcing requirements to push onto industry, EPA would act more as a leader in national debates about problems and priorities. Rather than just force industry to develop and install state-of-the-art technology, EPA would promote the development and dissemination of new technologies.

This conception of EPA's role is taking root. Yet EPA's role as chief national environmental regulator is still its most visible one. With a few exceptions, enabling laws still push it to set and enforce regulatory standards. How do agencies like EPA develop regulatory policy? What can this tell us about the agency and how it makes policy? These topics come next.

THE REGULATORY DECISION PROCESS

Through its first two decades, EPA's main role was to serve as the national environmental regulator. Regulations—ambient quality goals, effluent guidelines, pesticide registrations—were its most important product. Even when it did not set national standards, EPA had oversight and approval authority over standards that were set at the state level. EPA has been foremost a regulatory agency; to know its regulatory process is to know much about EPA itself.

Regulation is not for the faint-hearted. Much of the costs—in excess of $100 billion—of complying with environmental programs each year are driven by EPA rules. For environmental groups and others, the stringency of environmental regulations has become an important indicator of EPA's commitment to implementing the law. For those that bear the direct costs of regulation—industry and, increasingly, state and local governments—regulations affect planning and often financial survival. Regulatory decisions bring many groups together in an adversarial process with uncertain outcomes.[34]

The sources of EPA's rule making authority and of the constraints on its discretion come from the laws reviewed earlier. At any time, EPA is working on about two hundred fifty rules with important policy implications under these laws, in addition to hundreds of procedural and technical rules. Some rules are easily grouped, such as NSPS or effluent guidelines, both of which set technology-based standards for industrial categories. Other rules are one-of-a-kind, like location standards for waste treatment facilities or a rule banning uses of asbestos. EPA publishes a list of the rules currently being developed in its semiannual *Regulatory Agenda*.[35]

Designing a process for writing regulations in an agency like EPA is a challenge. EPA's internal process mirrors the external pressures and constraints that affect environmental policy making generally. Rule making occurs in an atmosphere of intense scrutiny, with participants from inside and outside of the government attempting to influence choices. Although the agency is organized to address problems along media lines, causes and solutions often cross those lines. Differences among laws reinforce these distinctions. Issues are complex, and scientific premises often are uncertain. Specialists from several areas must contribute for the agency to achieve sound and defensible policies. The political leadership comes and goes regularly. In sum, the conditions for making regulatory policy are less than ideal.

The internal process was designed with these conditions in mind. This means that it must encompass three goals. First, it must harness diverse specialties and organizational perspectives into a collective enterprise. The agency itself consists of several program directors and ten regions; each is organized around a programmatic, functional, or geographic set of responsibilities and brings varied perspectives to bear on issues. Add to this the interdisciplinary nature of most problems and their solutions. For example, issuing an air standard may require work from engineers, toxicologists, regional planners, statisticians, economists, and lawyers, among others.

Second, it must enable the political leadership to shape policy, to assure democratic accountability and control of the bureaucracy.

American public administration relies on the principle that a top layer of political appointees, named by the president, decides policy with the support of career civil servants. At EPA as elsewhere, the quality of political appointees has varied. Even for those who are capable managers and have substantive competence on the issues, the role the administrative system defines for them is not an easy one. Political appointees remain on the job for about two years, on average; this is barely enough time to get to know the people, master the issues, relate to the constituencies, and begin to shape the course of policy.

The third goal is to overcome the fragmentation in the institutional setting and to integrate regulatory responses across environmental media. Consider three situations. In one, the solution to an environmental problem creates others. Sewage treatment plants remove contaminants from water, but they create sludge containing lead, cadmium, and other toxics. In a second situation, the sources of a problem cross media lines, so solutions that are confined to one medium will be ineffective. Reducing groundwater contamination requires coordinated responses from surface water, hazardous waste, pesticides, and agricultural program managers. In the third situation, risks that are addressed through one program may be addressed more effectively through others. Lead poses very serious health risks, and exposures should be reduced when possible. The evidence on lead reveals, however, that exposures from the lead in paint chips, soil, and drinking water present more health risk than exposures from abandoned waste sites or from the application and disposal of sewage sludge. Having information on relative exposures and risks across environmental media enables policy makers to devise a more effective response. A well-designed internal process makes this integrated view possible.

So the internal process for developing regulations needs to meet each of these requirements: to harness diverse perspectives and specialties; to enable top policy officials to shape choices; and to anticipate and address cross-media issues. The following discussion describes EPA's management approach to meeting these requirements under three headings: priorities for regulatory development, participants in the internal process, and stages in regulatory development.[36]

The internal process also must accommodate requirements in the Administrative Procedure Act (APA).[37] Passed in 1946, its purpose was to limit administrative power and protect the rights of private parties affected by government action. It affects EPA rule making in several ways. It grants a right of judicial review to "any person suffering legal wrong because of agency action, or adversely affected or aggrieved by agency action." It requires agencies to publish rule making proposals in the *Federal Register* for public comment before issuing them as final, respond to public comments, and allow a minimum of thirty days before final rules become binding.[38] The APA defines the procedural framework for federal rule making.

PRIORITIES FOR REGULATORY DEVELOPMENT

The number and scope of regulatory issues on EPA's agenda is large. To treat each one as having equal importance would impose impossible demands on its staff, policy makers, and analytical resources. Deciding what is important and establishing priorities among issues is essential in effectively managing rule making.

EPA deals with regulations in roughly three categories. The highest priority category is a "major" rule as defined in Executive Order (E.O.) 12291. The order lists several criteria for what constitutes a major rule: an annual effect on the economy of $100 million or more, large increases in costs, or "significant adverse effects" on competition or employment.[39] Under the order, agencies must prepare a Regulatory Impact Analysis (RIA) that estimates the costs and benefits of major rules. Major rules undergo a longer OMB review period (30–60 days) than nonmajor ones (10 days). Within EPA, major rules receive more analysis and more management direction.

Major rules account for a small percentage of the rules that EPA issues. Others are classified as "significant" or "minor." Significant rules have national policy implications, impose economic burdens on industry or state and local governments, or affect health or environmental quality in important ways. Examples are effluent guidelines, toxic air standards, and rules governing permits for sources that discharge effluents into storm water sewers. This covers a large number of rules; EPA is working on more than two hundred significant rules at any time. Minor rules (delegations of program authority to states, procedural issues) account for hundreds of additional regulatory actions each year.

The basic provisions of the APA apply to all three classes of rules. Certainly minor rules generate less public comment, and the comment period itself is usually shorter. Within EPA, however, the three classes are treated differently. Major rules warrant an RIA, a full risk assessment, a longer OMB review, more policy debate, and interoffice participation. For most significant rules, there is an economic analysis, a risk assessment, and participation by offices across the agency. Minor rules are the subject of minimal participation and often are delegated to the regional or assistant administrator who has lead responsibility for the action. Risk or cost analysis usually is unnecessary for minor rules.

PARTICIPANTS IN THE INTERNAL PROCESS

There are several participants in the internal process. The lead program office (e.g., the Office of Water) directs the technical work, defines the issues, prepares the work plan, and manages the interoffice work group that develops the rule. For each significant rule, there is a staff work group with a representative from each office that is interested in the issue. The policy, legal, research, and enforcement offices are active in most work groups. Regional offices participate in developing rules that present implementation or enforcement issues or have special geographic importance. Other program offices (air, solid waste) take part in rules posing cross-media or generic policy issues.

Another set of participants is the political leadership. In the enabling legislation (such as the Clean Air Act or Clean Water Act), Congress delegates authority to the administrator, who then relies on the AAs and the career staff to exercise that authority. It is the political leadership—the administrator, the AAs, the regional administrators—who are accountable to Congress and the president. The administrator signs regulations, unless he or she delegates that authority to the AAs or the regional administrators.

Assistant and regional administrators may participate formally in a senior-level policy review, which comes near the end of the process. Here they officially concur or nonconcur with the rule as it has been drafted by the lead office and work group. Nonconcurrence may lead to a decision meeting with the administrator if offices are unable to resolve disagreements. This senior policy review is one of several points at which political appointees may shape policy decisions; others include early reviews of policy options, decision meetings arising from work group deliberations, and negotiations among AAs or with OMB or the White House.

The staff-level work groups under the charge of the lead office and the political leadership are two sets of participants in the internal process. But there are others. Several external advisory groups also counsel the agency on regulatory issues. The Science Advisory Board (SAB) offers advice on scientific issues and reviews the technical analyses underlying rules. The agency also relies on other technical advisory bodies, like the FIFRA Science Advisory Panel (SAP) or the National Air Pollution Technical Advisory Committee (NAPTAC). Other federal agencies are consulted on topics of mutual concern, such as the Department of Energy on utility issues or the Department of Agriculture on pesticide issues. And, of course, OMB reviews EPA rules.

STAGES IN REGULATORY DEVELOPMENT

The process of regulatory development is best understood as consisting of six stages: (1) start-up and planning; (2) issue development, options analysis, and drafting of the proposal; (3) review and publication of the proposed rule; (4) public comments and evaluation; (5) drafting and review of the final rule; and (6) promulgation and implementation. The length and significance of each stage varies; they depend on the scope and effects of the rule, its technical and analytical complexity, and the existence of statutory deadlines. Figure 2 presents the six stages and the typical range of time periods for significant and major rules.

Start-up and Planning The AA for the lead office initiates work on a rule, usually in response to a statutory requirement or a program need. The formal mechanism for starting work is a Start Action Notice (SAN), which notifies other offices and invites participation in the staff-level work group.

Issue Development, Options Analysis, and Drafting For rules of any importance, a major task is to define the issues, organize them, and pull together the data needed to resolve them. The range of possible issues is large. Some present a matter of basic strategy, such as whether to require the same technological controls for all sources of a toxic air pollutant (a technology-based approach) or to vary the sources' requirements according to their effects on health (a risk-based approach). Others are more technical, such as which monitoring methods to rely on or how much data to require from industry. Legal issues inevitably arise, from how to interpret provisions on economic analysis to determining what level of risk must be documented to successfully defend the rule in a court challenge.

It is also at this stage that key analyses are designed and drafted.

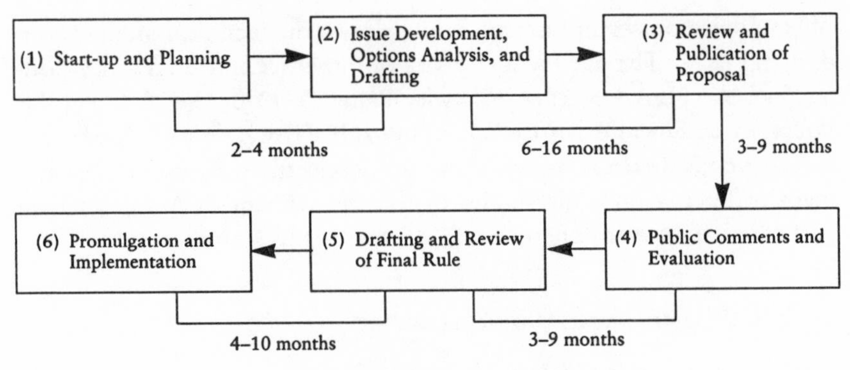

Figure 2. Typical stages in regulatory development (for significant and major rules).

Most significant and major rules are the subject of an economic impact or cost-benefit analysis. These analyses, in turn, often depend on the agency completing a technical analysis of control options and an assessment of health or other risks. In addition, the agency completes a Regulatory Flexibility Analysis, for rules that affect small businesses or governments, and an analysis of the information the rule requires from the public.

Review and Publication of the Proposal Moving from issues, options, and analyses to a fully drafted and documented proposed rule that reflects an agency consensus is no easy task. As the draft reaches its final stages, new technical, legal, or policy issues may come up which require analysis or decisions.

At some point, however, the lead office concludes that the work group has gone as far as it can, and the rule is slated for a "work group closure meeting." The purpose of this meeting is to determine whether the regulatory package is ready for AA-level review and for approval by the administrator. It is a final opportunity for all offices in the work group to raise questions, suggest changes in the rule or supporting analyses, and state what issues remain to be decided by senior management. After this, the lead office makes any necessary revisions in the proposed rule and enters it into the AA's policy review. The goal is to winnow the issues, to assure that only major ones are raised at the top and that major issues are in fact addressed at this level.

Once the agency has reached consensus on a proposal, it is submitted to OMB for review. Chapter 3 describes OMB's role in detail. After OMB's review and the resolution of any issues that OMB raises, the proposal is ready for the administrator's signature. The rule and the preamble are two of many documents making up the decision package, which also includes the regulatory or economic analysis, technical support documents (risk assessments or engineering studies), and comments from any AAs who want to offer advice to the administrator. The rule then is signed and published in the *Federal Register* for public comment.

Public Comments and Evaluation The APA requires that there be an opportunity for the public to comment on rules before they are issued as final. The formal mechanism for public comment is publication in the *Federal Register*; supporting documents are available in a docket that EPA establishes for each rule making. The APA sets thirty days as the minimum comment period, but agencies usually allow sixty to ninety days for important rules. The amount and content of public comment varies with the rule's importance and the controversy associated with it. Public comments serve several purposes. They give the agency an opportunity to hear reactions from industry, state officials, environmental groups, and others. Technical defects, high compliance costs, flaws in a scientific analysis—all can be brought to EPA's attention in the comments.

Drafting and Review of the Final Rule The pattern for developing a final rule after the comment period varies. If the comments are extensive and raise substantive issues, there may be more meetings with the political management and more analysis. At other times, responding to comments is a matter of explaining why the agency chose a particular option, justifying the rule further, or clarifying the rule's technical provisions. Internal and OMB review for a final rule follows a course similar to that for the proposed rule. After the administrator approves and signs the final rule, it appears in the *Federal Register*. Once it is published, the rule becomes legally binding on the parties who are covered by its requirements.

Promulgation and Implementation The APA requires at least thirty days warning before rules become binding, but agencies often allow more time. Of course, the final provisions will not surprise most in the regulated community, because there has been a period of data collection and public comment leading up to the final rule. However, rules that affect small firms may not become known to them, even after they have taken effect. For example, rules that apply to generators of small quantities of hazardous waste and owners of underground storage tanks cover hundreds of thousands of small firms, including many "mom and pop" businesses. Most are not in a position to retain legal or technical advisers on compliance, or to make the financial outlays necessary to comply quickly, if at all.

A near certainty for most important final rules is a court challenge. There simply is too much at stake for it not to be worthwhile for someone to invoke judicial review. The worst a party can do is lose, and the resources expended may be a small fraction of the compliance costs a large firm or industry would incur. The typical pattern in a court suit is for industry to claim that the regulatory requirements are too stringent and for environmental groups to claim that they are not stringent enough. Whether the agency wins or loses its case, the lawsuit can delay implementation of the regulation.

EVALUATING THE INTERNAL PROCESS

EPA's internal process has adapted over the years to the varied decision styles of its top management and to the changing demands from outside the agency. In the Costle years, for example, the emphasis was on reaching consensus among offices. Issues were elevated for decision by the administrator if they could not be negotiated by middle management or the AAs. Under Ruckelshaus and his deputy, Alvin Alm, there was more top-level involvement in regulatory decisions, especially on Alm's part. An astute analyst and advocate of rational decision making on regulatory issues, Alm oversaw the creation of an "options selection" process that identified the major regulatory issues and brought them before political managers for debate and decision at an early stage, while there was time to affect choices. Later, the agency moved back to an operating style that elevated issues only when they could not be resolved at lower levels. This meant that issues often reached the administrator late in the process, if at all.

Many pressures from outside the agency affect rule making as it occurs inside the agency. One has been Congress's penchant to load EPA's reauthorized laws with "mandated" actions.[40] These are actions in which Congress gives the agency little or no discretion. Often, the laws turn into a lengthy "to do" list for EPA. The Water Quality Act of 1987, for example, directed EPA to issue twenty-five rules, develop forty policy documents, prepare thirty-one studies or reports to Congress, begin fifty implementation actions, and meet ninety deadlines by 1990. Mandated actions—often taken under pressure of statutory deadlines—dominate EPA's regulatory agenda. They increase the sheer number of rules that must be issued while limiting the time available to issue them. EPA has had to adapt at times by cutting back on its analysis of policy options, limiting internal policy reviews, and consulting less widely with outside groups.

Another source of external pressure is OMB. Under Presidents Reagan and Bush, OMB was a source of controversy and delay. OMB's interests in limiting costs and its critical view of EPA's programs affected the internal dynamics of rule making. This may change, but OMB review in the 1980s and early 1990s greatly reduced the control that agencies like EPA had over analyses and policy decisions.

How well does the process meet the goals? The work groups, middle management oversight, and AA review meet the minimal need to bring diverse offices and experts together. The success of the process in allowing for top-level direction of policy has waxed and waned under different leadership styles; the Ruckelshaus and Alm period was a high point in achieving administrator and AA-level involvement in decisions.[41] Efforts to further integrate policy have gone through several stages. The work group process and policy reviews usually achieve some level of *coordination* among different parts of the agency but often do not achieve a more comprehensive *integration* across environmental media and offices. A step toward achieving better coordination and integration is the effort to cluster rules that affect the same industry, pollutant, or resource.[42]

IS THERE A BETTER WAY? REGULATORY NEGOTIATION

The rule making process that has evolved over the last several years exhibits both strengths and weaknesses. One strength is that it grants some due process to affected parties, with guarantees of participation and judicial review. The close scrutiny from private parties as well as review by the courts may promote decision making that is well reasoned, supported by the facts, and consistent with the law. As a result, rule making generally is principled, orderly, and predictable.

Rule making can also be slow, adversarial, and contentious. The requirement that regulations have a rational basis, with factual conclusions clearly supported, imposes high information demands on regulators. Documenting the need for environmental decisions is especially demanding, because of the scientific uncertainty about risk. Different sides bring their scientists to the debate over evidence, usually to weaken the arguments of the opposition. The inevitability of judicial review prolongs decision making and puts the parties on an adversarial footing from the start. Judges often must end up making decisions that Congress delegated to the agency.[43] Is there a better way?

One idea that EPA and other agencies have explored over the last decade is regulatory negotiation, which applies the techniques of alternative dispute settlement (ADR) to rule making. The heart of the regulatory negotiation process is to identify interests with a stake in a proposed rule, get them to agree to join a negotiating committee and attempt to resolve the issues, and then help them come up with a proposed rule that is acceptable to everyone on the committee and presumably to all the groups affected by the proposed rule.

These negotiations share three characteristics.[44] First, they complement but do not replace the conventional rule making process. The negotiation process conforms with the Administrative Procedure Act. The aim of the negotiations is to produce a proposed rule that reflects a consensus of the parties and is published in the *Federal Register* for public comment. Second, EPA participates as a "party at interest" in the negotiations. It also commits to publishing the committee's consensus as a proposed rule, as long as the proposed rule that emerges from the negotiations is consistent with the statute. Third, once it is constituted, the negotiating committee has control over its own operations, composition, use of resources, and the terms of its dissolution. The negotiations usually occur in several two- to three-day sessions spread over six to eight months.[45]

EPA's experience with regulatory negotiation so far suggests that it has several advantages over conventional rule making. The participants appreciate the chance to deal directly with the people who represent varied interests. The process appears to be useful in drawing out practical information about regulatory issues from the parties. There is an incentive for manufacturers to explain, for example, why implementing a provision would be difficult for them, or for community activists to explain why they might be suspicious of streamlining a process to issue permits for waste facilities. Outside parties appreciate the opportunity to discuss their issues with senior agency officials at a negotiating table. Finally, a negotiated rule is likely to find greater acceptance by the affected parties and to encounter less chance of litigation.

Despite its merits, regulatory negotiation should be used selectively. Some issues are so big and complex that it would be difficult to assemble a workable committee. A negotiation may be inappropriate when there are major scientific uncertainties or policy issues regarding public health or ecological resources. The evidence suggests that regulatory negotiation as a policy process is best suited to rules that involve implementation or procedural questions rather than fundamental questions of values. Viewed in this way, it offers a useful complement to conventional rule making.

LAWS, AGENCIES, AND RULE MAKING

Would this country be better off with one organic statute that defines goals and authorities consistently across all environmental programs? Should cost and benefit issues be evaluated the same way in air quality and hazardous waste, so that any trade-offs made are consistent across programs? Should the preference for command-and-control regulation embodied in our environmental laws be changed, to allow more room to use cooperative approaches and economic incentives? Does EPA's current organization merely reflect the fragmentation in the statutes and the broader political environment rather than provide a basis for more integrated policy making? Is the rule making process in its current form too adversarial?

These are only a few of the many questions that a review of environmental laws, EPA, and the rule making process will raise. Certainly a "yes" to any of these questions is defensible from the standpoint of rational policy. Having so many laws to deal with such a diverse set of interrelated problems complicates policy making. The fact that ambiguity about cost and benefit issues has not been clarified across laws and over several reauthorization cycles contributes to continued analytical confusion. Although it is hard to say what form of organization would make more sense for EPA in the current political environment, it is fair to say that its origins as a patchwork agency create continuing obstacles to the integration of policy and to EPA's ability to act strategically based on risk. The rule making process has so few defenders that it is almost axiomatic in Washington that we should look for better ways.

But the institutional picture is far more complicated even than the picture presented in this chapter. The enabling laws, EPA, and the rule making process are only part of the institutional landscape for environmental policy making. Congress passes laws, then watches over their implementation like a nervous parent taking children through a crowded amusement park. An "environmental president" begins his term by declaring that there should be "no net loss" of the nation's wetlands and signing the most expensive environmental regulatory law in U.S. history (the CAAA of 1990), then ends it by dropping his support for wetlands and declaring a moratorium on new regulations. A rule making process goes on for three or four years, only to end up in court, where a judge may tell the agency to start over. Other federal agencies administer programs that have as much of an effect on the environment as EPA, yet may not share EPA's goals.

The next chapter turns to the rest of the institutional picture—Congress, the White House and executive branch, the courts, intergovernmental and nongovernmental players—to complete the discussion of the institutional setting for policy.

CHAPTER THREE

Institutions II

As the presidential campaign began heating up in the winter of 1992, President Bush took one of many steps marking his reelection bid. He directed federal regulatory agencies to conduct a ninety-day review of rules and programs "to identify and accelerate action on initiatives that will eliminate any unnecessary regulatory burden, or otherwise promote economic growth." He declared a moratorium on the issuance of new regulations for ninety days. The purpose of the review and moratorium was to "weed out unnecessary and burdensome government regulations, which impose needless costs on consumers and substantially impede economic growth."[1] This moratorium was later extended 120 days, then past the election. Except for rules with court or statutory deadlines or rules that were deregulatory, no new rules were issued for the rest of the Bush presidency.

It was no secret within the administration or anywhere else that EPA was a main target of this order. In the nearly twenty-two years of its existence, EPA had accumulated authority under more than a dozen major laws and had issued a large body of regulations affecting nearly every aspect of American business and society. Environmental protection cost the United States $110 billion a year by 1992, well over 2 percent of the gross national product (GNP). Environmental regulations affected the air Americans breathed, the water they drank and swam in, the chemicals used in industry and in the home, the pesticides applied to food, where and how wastes were disposed of—even the life and death of wetlands and estuaries. That the environment was better off than it would have been without the national program was obvious to most people. But at what cost?

The regulatory moratorium was the latest event in over two decades of struggle between environmental and economic values. The cast of characters may have changed over the years, but the central tensions in the debate have been fairly constant. At a societal level, the debate has been between those who value economic growth and development, on one side, and those who advocate environmental quality and conservation, on the other. Advocates of growth argue that the growing costs of environmental programs divert resources from investment and production and leave the country worse off economically. Advocates of environmental quality argue that economic growth on its own will not improve but lessen the quality of life if its effects on the environment are not accounted for.

At a more specific level, we see tensions between growth and the environment all the time: Do we allow a factory to expand and increase air emissions in an area that already exceeds air quality standards? Should we cancel uses of a pesticide that poses small health risks when farmers depend on it and there are no effective substitutes? Should developers be allowed to build marinas when it means draining and covering acres of wetlands? The moratorium was simply playing out nationally the more specific conflicts that are occurring across the nation and around the world all the time.

The moratorium was only one event, an election year event at that, when any event has to be discounted by the political needs and symbols of the moment. In the decades-long struggle between the environment and economics, President Bush was simply casting his lot with the side that seemed to offer the greatest political benefit at the time. But there was a consistent theme behind the moratorium and what it represented, a theme much broader than the ideological issues dividing the parties. Public officials act on the basis of institutional perspectives as well as on the basis of party or ideological differences. Events such as the moratorium reflect as well as shape these institutional perspectives. They reflect them because they are one of many weapons a president has to counter the influence of Congress, the courts, and interest groups. And yet the moratorium was constrained by institutional influences—by the directions of the enabling laws, by the authority of the courts to review agency action (or inaction), and by interest groups' use of Congress, the courts, and other institutional levers. The tension between the environment and economics was not the only tension in policy making, but it was a central one.

These and other tensions in environmental policy making are defined and sometimes resolved in the context of institutions and laws. In this chapter, we continue our look at institutions and their role in environmental policy making by turning to the other participants: Congress, OMB and the executive branch, other federal agencies, the federal courts, state and international institutions, nongovernmental forces, and citizens.

CONGRESS AND LEGISLATIVE OVERSIGHT

Asked to pick the one institution that has been the most important initiator of national environmental policy over the last few decades, most observers would select Congress. Unlike some policy areas, where there is a pattern of executive leadership and legislative response, the source of change in environmental policy has usually been Capitol Hill. In the remarkable series of laws that began with the NEPA and continues in the cycles of revision of the RCRA, the CAA, and other laws, Congress has taken the lead. As the laws emerged from the Hill, they became more and more prescriptive and brought closer and more critical oversight from Congress.

Because Congress has spawned most environmental laws, its members exhibit a certain protectiveness toward their progeny. To the key committee and subcommittee chairs, these are "*their* laws," and "statutory intent means *their* intent."[2] This leaves little room for agency discretion or experimentation. Even when statutes were unclear, Senators Edmund Muskie and Robert Stafford or Representatives James Florio and Henry Waxman were eager to share their interpretations of what EPA should do. To its defenders on the Hill, close oversight is an asset to EPA, a counterweight to the power of the Office of Management and Budget and large agencies like the Departments of Defense and Energy. To critics of legislative oversight as practiced recently, Congress attempts to "micromanage" agencies and provides little constructive or responsible legislative oversight.[3]

Congress influences environmental policy in several ways, one of which is the enabling laws discussed in chapter 2. The environmental statutes rely on a variety of strategies and mechanisms for setting goals and directing how they are to be carried out. These statutes incorporate a mix of technology-based, risk-based, and cost-based approaches to goal and standard setting. Congress at times has taken on the job of setting national goals and at other times has left it to states. Some statutes leave costs out of the decision equation entirely; others explicitly direct EPA to balance costs with benefits in justifying a decision. One law is designed to reduce pollution in a given medium, like ambient air or drinking water; another focuses on the sources or effects of pollution, such as pesticides or acid rain. Although the air program relies almost entirely on the private sector to fund environmental controls, the water program includes a large share of public funding, in the form of construction grants (now a state revolving fund) to fund sewage treatment plants. In sum, anyone seeking consistent strategies across problems and programs will come up empty. But Congress's influence over policy begins with the enabling laws.

There are several mechanisms allowing Congress to oversee agencies as they implement their enabling laws. These include hearings, reports to Congress, letters from members of Congress asking for information or answers to questions, the appropriations process, Senate confirmation of presidential appointees, and audits by the General Accounting Office. Let us look at the structure for congressional oversight before discussing some of them.

THE STRUCTURE OF CONGRESSIONAL OVERSIGHT

The process for making national environmental policy in this country is fragmented. Any search for the causes of fragmentation must begin with the structure of congressional oversight. In each house, several committees and subcommittees share authority over EPA's programs. Some have jurisdiction based on a specific subject area; others have broad "oversight and investigations" authority.

The single most important entity is the Senate Committee on the Environment and Public Works, which has authority over all but the pesticides program. Chairing this committee almost guarantees a senator a major role in policy. Senator Muskie used his position as chair to lead Congress through passage of the major statutes in the early 1970s. Jurisdiction is divided among four subcommittees: Superfund, Recycling, and Solid Waste Management; Toxic Substances, Research, and Development; Clean Air and Nuclear Regulation; and Clean Water, Fisheries, and Wildlife. Again, the position of chair of each of these subcommittees offers a visible platform for leadership. Sen. George Mitchell of Maine used the chair of the Environmental Protection Subcommittee (under the previous subcommittee structure) to reauthorize the Clean Air Act in 1990. Sen. Frank Lautenberg used the chair of the Superfund and Oversight Subcommittee (also under the previous structure) to take a leading role on hazardous waste cleanup, a major issue with his New Jersey constituency.

Although Environment and Public Works is the key committee in the Senate on environmental issues, others oversee specific issues. The Committee on Agriculture, Nutrition, and Forestry has authority over the FIFRA and pesticides. In both houses of Congress, the fact that the agriculture committees oversee pesticides policy has major implications. They tend to be more oriented toward farmer and grower interests than the environmental effects of pesticide use. The Water and Power Subcommittee of the Senate Committee on Energy and Natural Resources oversees groundwater policies.

The situation in the House is more complicated; no single committee has jurisdiction as broad as that of Environment and Public Works in the Senate. Five House committees principally oversee EPA programs. The Energy and Commerce Committee has Subcommittees on Health and the Environment and on Transportation and Hazardous Materials. The Public Works and Transportation Committee (its Subcommittee on Water Resources and Environment) and the Merchant Marine and Fisheries Committee (its Subcommittee on Environment and Natural Resources) oversee water pollution issues. The Agriculture Committee holds authority over pesticides matters. The Committee on Science, Space, and Technology has jurisdiction over environmental research and biotechnology issues.

Overall, Energy and Commerce is probably the most influential of the House committees, especially under the strong leadership of Chairman John Dingell of Michigan. It handles air, toxics, waste, drinking water, and groundwater. On some issues, oversight in the House gets very complicated. For example, three committees share authority over groundwater; a fourth (Agriculture) has authority if groundwater contamination by pesticides is involved.

Besides the major oversight committees, many others play a role on specific issues. For example, the Senate Governmental Affairs and House Governmental Operations committees have broad jurisdiction over all issues, including the environment. And the Appropriations committees in both houses have budget authority over EPA and environmental programs. Table 2 lists the principal committees and subcommittees that oversee EPA and environmental programs and their jurisdiction, as of 1993.

There are many cooks stirring the congressional oversight pot; this is one of several factors that fragment the policy process.

TABLE 2 PRINCIPAL CONGRESSIONAL OVERSIGHT COMMITTEES AND SUBCOMMITTEES (1993)

Committee/Subcommittee	*Jurisdiction/Issues*
Senate	
Environment and Public Works	
Subcommittees:	
Superfund, Recycling, Solid Waste Management	RCRA, CERCLA, federal facilities
Toxic Substances, Research and Development	TSCA, NEPA, research
Clean Water, Fisheries, and Wildlife	CWA, SDWA, ocean dumping
Clean Air and Nuclear Regulation	CAA, indoor air, nuclear
Agriculture, Nutrition, and Forestry	
Subcommittee on Agricultural Research, Conservation, Forestry, and General Legislation	FIFRA, pesticides, forestry
Energy and Natural Resources	
Subcommittee on Water and Power	Groundwater, irrigation
Appropriations	
Subcommittee on VA, HUD, and Independent Agencies	EPA budget
House	
Energy and Commerce	
Subcommittees:	
Health and the Environment	CAA, environment, SDWA, FFDCA
Transportation and Hazardous Materials	RCRA, TSCA, CERCLA
Merchant Marine and Fisheries	
Subcommittee on Environment and Natural Resources	Coastal, marine, endangered species, NEPA
Public Works and Transportation	
Subcommittee on Water Resources and Environment	Water pollution control, CWA
Agriculture	
Subcommittees:	FIFRA, pesticides
Department Operations and Nutrition	
Environment, Credit, and Rural Development	
Science, Space, and Technology	
Subcommittee on Technology, Environment, and Aviation	Environmental research, biotechnology
Appropriations	
Subcommittee on VA, HUD, and Independent Agencies	EPA budget

MECHANISMS FOR CONGRESSIONAL OVERSIGHT

Congressional oversight of environmental policy warrants a book of its own. There are a variety of mechanisms available to the people on Capitol Hill for making their views known to EPA and for influencing environmental decisions. Some draw relatively little public attention. For example, each statute routinely includes a list of required reports to Congress on several issues. Members of Congress use the General Accounting Office—Congress's auditing and evaluation arm—to study issues and offer recommendations. Letters to EPA from members of Congress serve many functions: answering routine requests from constituents, chastising the agency for decisions it has made (or failed to make), stressing a member's preference on a pending issue. In the mid-1980s, for example, EPA received four thousand to five thousand letters a year from Congress.

Hearings are perhaps the most visible of the oversight tools. Because so many committees and subcommittees have jurisdiction over EPA, and because its issues draw public interest, EPA officials often have to testify before congressional committees. Over the three-year period 1984–1986, EPA officials testified on about 200 different occasions. About half of these appearances dealt with specific issues—asbestos, acid rain, and so on. Others concerned statutory reauthorizations, the budget, confirmation hearings, and other topics.[4] In comparison, in the two-year period 1989–1990, EPA officials appeared before congressional committees 245 times (132 in 1989 and 113 in 1990), almost twice the rate of the earlier period. Administrator Reilly testified on 28 of these 245 occasions. In the first five months of the Clinton administration (through the end of May 1993), EPA officials testified in front of Congress 44 times; 13 of these involved Administrator Browner.[5]

My own experience is that political appointees regard an invitation (often a demand) to testify before a probably hostile committee or subcommittee with dismay, akin to a political root canal. I have also heard stories of an agency official sitting through an unpleasant tongue-lashing, only to have a member or staffer come up after the hearing, exchange friendly greetings, and offer the encouragement that things were really going pretty well, the grilling was just to make points for the record or the media.

The appropriations process is another effective tool. Money talks, here as elsewhere. A lack of responsiveness to a member with a key appropriations post can have consequences later. The needs and interests of members with budgetary clout affect where money in the EPA budget goes. For example, because Rep. Bob Traxler chaired a House appropriations subcommittee, an EPA laboratory was located in Bay City, Michigan, his home district. In addition, Congress often attaches riders to appropriations bills that direct EPA to conduct a study, change policy, or take some other action. In 1990, for example, the House directed EPA to move ahead more aggressively on its pollution prevention program and outlined specific steps it should take. In 1992, a Senate rider directed EPA to study small communities' ability to pay for the treatment technology required by the proposed drinking water standards for radionuclides.

A fragmented structure, the sense of ownership of specific laws and issues, a need to claim credit back home for aggressively protecting constituent interests (either to protect health or jobs), a genuine interest in solving a problem (whether hazardous waste, drinking water, or urban air) that is of special concern, divided control of the White House and Congress—all of these affect the oversight process. It was no surprise when the National Academy of Public Administration concluded in 1988 that "incentives for episodic, ad hoc oversight, are generally far stronger than those that encourage comprehensive, systematic evaluations of statutes and programs."[6]

ASSESSING CONGRESSIONAL OVERSIGHT

What can we say about the effects and merits of legislative oversight? It should be noted first that legislative oversight is the foundation of democratic accountability in the administrative state. Whatever power and legitimacy agencies have is due to the authority delegated to them by Congress. In the early years, the courts were concerned that when authority was granted to agencies, it should be limited, under the delegation doctrine. Since then, the courts have focused more on other issues, and the scope of the allowable delegations to agencies has become broader. Yet Congress has a constitutional responsibility not only to pass the laws but also to oversee their implementation by administrative agencies. Both the growth of a complex committee structure and the tendency for legislators to specialize are reactions to the need for oversight of technical issues and expert agencies.

As for the effects of the close congressional oversight of environmental policy, the answer has to be much more complicated. Critics would level several charges at Congress. They would say first that Congress (through its committees and subcommittees) has acted as a micromanager of policy that ignores the big picture. They would observe that members use positions of authority (such as committee chairs) to push their own agendas and draw attention to their own issues. They would point to a tendency by Congress to dramatize problems, call for solutions from EPA or other agencies, impose strict (usually unachievable) requirements for action on those agencies, and then criticize them for failing to eliminate the problem. At the same time, critics would say, Congress fails to think through just what it would take to address problems, to allocate anything close to the resources needed to solve them, or to provide the political support necessary to implement programs and impose sanctions. Congress, the critics would say, is eager to reap the political benefits of aggressively dealing with problems but is much less eager to bear the costs of the solutions.

Defenders of Congress would paint a different picture. They would draw attention to the problems EPA and other agencies have had in issuing rules on toxic air emissions or achieving the goals of the Superfund cleanup program. They would point out that in the first years of the Reagan administration, EPA was reluctant to set tough standards and enforce pollution laws. For environmental groups, Congress has been an ally, offsetting the White House's reluctance to move more aggressively on problems of hazardous waste, global warming, and acid rain. Strong political support for environmental programs has enabled Congress to act as a kind of conscience for EPA and other agencies—asserting public concern and holding EPA accountable for responding to it.

THE WHITE HOUSE AND THE EXECUTIVE BRANCH

We all know that Congress makes the laws and the executive branch carries them out. We should also know that beneath such a simple, civics book statement lie several layers of complexity.

Congress consists of a collection of specialized committees and subcommittees, a diversity of individuals seeking to do good for the environment or do well in their home districts, and a fragmented oversight process that pulls an agency first one way and then another. The executive apparatus is just as complex. As a technical matter, Congress delegates authority to the administrator to act to protect the environment. But administrators are selected by the president and depend on the White House for political and budgetary support, not to mention their jobs. Nor is the White House content to leave as vital an issue as the environment (and thus the economy, growth, and other social ends) to an agency acting under the direction of Congress. So the White House exerts its own influence on EPA and others who deal with the environment in one way or another. Making policy in the executive branch is as byzantine and complex as it is in Congress.

MECHANISMS FOR EXECUTIVE OFFICE CONTROL OVER POLICY

Appointments and budget are two standard mechanisms of control that enable the White House and the executive office to shape policy. By appointing senior-level officials, presidents can select like-minded people who are invested in the administration's goals and beholden to some degree to the White House for their futures. In the early 1970s, President Nixon pursued an administrative strategy for policy control that relied heavily on appointments.[7] When Presidents Gerald Ford and Jimmy Carter set out to deregulate the airline and trucking industries in the 1970s, their appointment of like-minded agency heads to regulatory commissions helped move the process along.[8] By appointing Anne Gorsuch administrator of EPA and a group of AAs who were sympathetic to industry's views on regulation in 1981, President Reagan paved the way for his administrative strategy for recasting EPA.[9]

Another mechanism for White House control of agencies is OMB's control over the budget process. Each year, the administration submits its budget to Congress. Congress makes the final budget decisions, but the president's budget is an important statement of the executive branch's priorities. An administration can influence policy through the budget process when other measures fail. In the early 1980s, for example, the Reagan administration used the budget process to greatly reduce EPA's capabilities. It took years for the agency to restore its resource base and staffing after these cuts.[10] Some argue that it has not yet recovered.

The budget that the administration submits to Congress in January of each year (for appropriations the next fiscal year, starting in October) is the product of a long process of analysis and debate in the agencies and between the agencies and OMB. EPA's process for preparing the OMB budget begins early in the year, usually with an annual planning meeting of the top leadership. The result of this meeting is a statement of themes and priorities that guide the preparation of detailed budget requests by assistant and regional administrators. In the summer, EPA submits a formal budget request to OMB. This and the negotiations surrounding it usually fit an incremental budgeting model: the agency defends the activities it will carry on in its base and justifies in more detail the resources it requests above the base (the increment).[11] In the OMB hearings, which are usually held in September, budget examiners review the budget request in detail. After an OMB passback to the agency and any agency appeals—in theory, to the president if necessary—the final administration budget is prepared and sent to Congress. EPA's budget is part of this larger budget.

OMB's role in clearing legislative proposals and comments on proposed legislation gives it additional control. Before any agency can comment on proposed legislation in its subject area, or propose legislative changes in its own enabling laws, it must clear those comments or proposals through OMB and defer to the administration's position on an issue. This means, for example, that any changes in the Superfund law or the Resource Conservation and Recovery Act—two statutes that are up for reauthorization in the 1990s—could be prepared under the direction of OMB and the White House.

Yet another mechanism for oversight by the executive office is OMB's review of rules.[12] OMB's regulatory review expands on its traditional role as budget manager and critic. It is also part of a shift from an old OMB of neutral competence and professionalism to a new OMB—one wearing the ideological stripes of the sitting administration.

OMB review has been part of EPA's life from the start. Only the form has changed. In 1971, OMB instituted a Quality of Life review of environmental, consumer protection, health, and safety regulations.[13] Since then, each administration has set out its own requirements for regulatory analysis and review in a series of executive orders. In 1974, for example, President Ford issued an order requiring agencies to prepare Inflation Impact Statements that estimated the costs, benefits, and inflationary impacts of major rules.[14] President Carter's E.O. 12044, issued in March 1978, was a broad effort to introduce more rational analysis and decision making in federal regulations. It directed agencies to assess policy alternatives, estimate the economic consequences of their decisions, complete a regulatory analysis for all major rules, write rules in plain English, minimize paperwork burdens, and prepare plans to evaluate rules once they were in place.[15]

Carter also created the Regulatory Analysis Review Group (RARG), which selected major rules for critical analysis by OMB and the Council on Wage and Price Stability (COWPS). RARG was a source of economic expertise that could second-guess agencies. Yet its review occurred as part of the public comment period; its approval was not a condition for issuing rules. It advised the president, but it did not clear policy through a rule-by-rule review.

The Reagan team pursued its deregulatory agenda with a Regulatory Relief Task Force under the leadership of then-Vice President George Bush and also staffed by OMB. It targeted several rules (at EPA and elsewhere) for revision that would reduce costs and burdens—a hit list of rules that industry was eager to change.

By far the most ambitious effort to exert executive office control over agency policies was E.O. 12291, issued in 1981. With this order, OMB review gained weight and grew teeth. The order included both procedural and analytical requirements. Agencies had to submit all rules to OMB's Office of Information and Regulatory Affairs (OIRA) for review before issuing them. OMB had ten days to review nonmajor rules, sixty days for proposed major rules, and thirty days for final major rules, but it could unilaterally extend these periods. The order required agencies to prepare an RIA for major rules—defined as those with an annual economic effect of over $100 million (or meeting other criteria). Within the constraints of their enabling laws, agencies were to issue rules only when the benefits exceeded the costs; they had to show that a rule would yield net benefits to society, or at least that they had chosen the most cost-effective way to achieve the agency's goal.[16]

Overall, EPA rules have constituted less than a tenth of the total rules OMB reviews in a year. (See fig. 3 for a breakdown by federal agencies for 1990.) Yet EPA's rules tend to be among the more important ones that OMB reviews. Of the eighty-two major rules that OMB evaluated in 1990, over 25 percent of them (21) were EPA rules.

What were the effects of E.O. 12291 on EPA? One was delay. Many rules pass through OMB review with no difficulty, but others are delayed when OMB raises issues. In 1990, for example, OMB reviewed 173 EPA rules, of which 21 were major.[17] About half of these were held by OMB past the initial review period, making it necessary for EPA to negotiate with OMB before issuing the rule.

We can get a better idea of the effects of OMB's oversight by looking at an EPA status report from October 1992 as an example.[18] Over a two-week period (October 6–19), EPA submitted nine rules to OMB (5 air rules, 2 toxics, and 2 water). In addition to these rules, two oth-

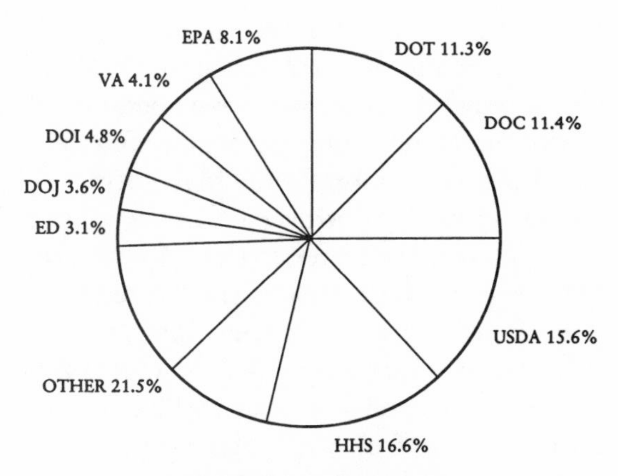

Abbreviations: DOT (Department of Transportation)
DOC (Department of Commerce)
USDA (Department of Agriculture)
HHS (Department of Health and Human Services)
ED (Department of Education)
DOJ (Department of Justice)
DOI (Department of the Interior)
VA (Department of Veteran's Affairs)

Figure 3. Total reviews by agency under E.O. 12291 (1990). Source: *Regulatory Program of the United States Government, April 1, 1991–March 31, 1992* (Washington, D.C.: Office of Management and Budget, 1991).

ers—on land disposal standards for hazardous wastes and water quality guidance for the Great Lakes—were at OMB, but these were still within the original time periods for review. What is more revealing is the list of rules for which OMB review had exceeded the time period specified in E.O. 12291. Fifty rules fell into this category. Half were air rules (reflecting the large number of rule makings under the 1990 CAAA). Most of the others were waste, pesticides, and toxics rules. Some had been at OMB for a long time. The winner was a major rule on groundwater standards for inactive uranium mill tailings; it had been under OMB review for forty-three months. More typical were rules on acid rain nitrous oxide (NOx) emissions (6 months); state water quality toxics standards (6 months); and location standards for waste facilities (1 month).

Because these rules were caught in the moratorium, they may have exaggerated the extent to which OMB delays rule making. But in fact, these cases were not atypical. In 1991, nearly 80 percent of the rules that were sent to OMB were delayed past the period specified in E.O. 12291. The total number of rules that were delayed almost tripled between 1990 and 1991, from 50 to 142. Looking at it from the other side, only 20 percent of the rules submitted in 1991 were cleared on time, compared to nearly 70 percent the year before.[19] OMB appears to have anticipated the moratorium; the number of rules it delayed increased in the year before the moratorium took effect.

When issues are especially controversial and agencies are unable to agree with OMB, the issue is elevated to the political level at the White House. In 1991, President Bush created the White House Council on Competitiveness, chaired by Vice President Dan Quayle, to act as an appeals board for agencies that were unhappy with OMB's decisions.[20] Other members of the council were the OMB director, the attorney general, White House Chief of Staff John Sununu, the chairman of the Council of Economic Advisors, and the secretaries of Treasury and Commerce. The council quickly became controversial because of its proindustry bias and power to change regulatory decisions outside the view of the public and Congress. The council and later the moratorium on new regulations cast a long shadow over the final year of the Bush administration. Both showed that it would push economics to the detriment of the environment in the reelection campaign—even to the point of setting up false dichotomies between the two.

One of the more visible and contentious issues taken to the council was a decision in August 1991 to relax the criteria for defining wetlands. EPA, Agriculture, Interior, and the Corps of Engineers had agreed on a wetland delineation manual that served as a basis for defining wetlands around the country. Pressure from farmers and others who were concerned about limits on their ability to cultivate or develop lands led the council to review the criteria, with an eye toward relaxing them; if they were relaxed, fewer acres would be defined as wetlands and more acreage would be available for development.

Usually, when four federal agencies agree on a policy, it is accepted by the White House. This time, the council intervened. Pressed for a relaxation in the criteria, the EPA administrator proposed a compromise that relaxed the criteria somewhat but retained EPA's authority to veto federal permits for developing wetlands areas. The council's intervention in the agreement led *The Washington Post* to observe, "The struggle over who should control environmental policy—the experts throughout government or the political and economic advisors to the president—is as old as this administration." But many wetlands experts complained, as one did, that "even for a White House whose top officials get involved in the smallest domestic policy spats, the resolution of this issue reached new heights of political intervention."[21] The wetlands criteria dealt with a number of narrow technical issues, like the number of days of soil saturation each year and its depth or the plant species found in wetland habitats. At one point, *The Washington Post* noted, the participants "had to be given a special wetlands glossary" for a decisive meeting.

Rarely have technical and political factors come together in this way on such a controversial issue. The wetlands controversy illustrates the degree to which the White House will intervene in decisions that appear to be narrow and technical but in fact have major political and economic implications. The Competitiveness Council gave the Bush White House yet another mechanism for influencing environmental policy in the administration's final year.

ASSESSING OMB AND EXECUTIVE OFFICE OVERSIGHT

OMB's regulatory review is a fixed and prominent piece of the institutional landscape.[22] What can we say about its effects on EPA and environmental policy? Let us begin, as few people do, with some positives. It is fair to say that only the president and the executive office are in a position to look broadly at regulatory policy. Agency heads may try to integrate across agencies and their jurisdictions, and they may succeed to some degree, but the process for making policy inevitably will be fragmented. Laws define and reinforce this fragmentation, as do the structures and incentives for legislative oversight. The White House and its OMB agents have a view of policy that few others can match. There is also something to be said from an accountability standpoint for giving the president the capacity to evaluate agency decisions. Congress delegates authority to the EPA administrator as a formal matter, but it is hard to argue that the president should not be able to influence how it is carried out. It is the president, not the head of EPA, who is accountable to the electorate.

Yet there is another side to the ledger. OMB has not lacked critics. One criticism is that OMB review interferes improperly with the delegation of power to agencies. Only Congress has the constitutional power to grant agencies the authority to regulate. The president has a constitutional obligation to "take care" that the laws are faithfully executed. This is interpreted as allowing the president to "supervise" but not to "displace" an agency's exercise of its congressionally delegated authority. Yet the lines between supervision and displacement are not clear, especially when Congress delegates more regulatory authority to agencies than the White House would like. The exercise of a president's supervisory power raises the danger of "displacement of authority vested in the relevant agency head."[23]

Another criticism is that OMB lacks the resources to second-guess an agency's technical expertise. There certainly is merit to this argument; compare the half dozen or so OMB staff assigned to reviewing EPA rules at OMB to the technical, policy, and legal resources available to the agency itself.[24] OMB responds that it is examining the validity of policy positions and the economic effects of rules, not the technical analyses behind them, and by focusing its priorities it can do the job. Yet as we have seen, delay is one of the consequences of OMB's rule-by-rule review. This delay is in part due to the backlog in reviews and the need for OMB staff to raise many questions about technical issues.

Whatever merit these criticisms may have, the larger part of the case against OMB review is based on how it is conducted, not the review itself. Although E.O. 12291 was carefully written to avoid giving OMB the power to reject or delay agency rules, in practice the order gave OMB virtual veto power. Once OMB staff raise an issue, agencies usually have not been able to move forward until the issue is resolved, if necessary by elevating it to the political leadership for resolution. There is also substantial criticism of the way OMB manages its review process. Midlevel staff at OMB have had authority to raise objections and delay rules that the AAs or the administrator at EPA have endorsed. OMB is often slow to define issues, to elevate them for management attention, and to release rules once their initial objections have been addressed.[25]

What is clear about OMB is that it is a major player in the institutional conflict between Congress and the White House over the control of environmental policy. It is also a large part of the reason that Congress has reduced EPA's discretion with highly prescriptive enabling laws. From the perspectives of the committee and subcommittee chairs on Capitol Hill, Congress delegates the authority to agencies, which they are expected to exercise under congressional supervision with minimal White House interference. So long as OMB's role was largely advisory, as it was under the Carter executive order, or was limited to long-established levers like budget and legislative clearance, Congress could accept it as a part of the normal pulling and tugging that characterizes executive–legislative branch relations. Once OMB acquired the authority to perform rule-by-rule review and clearance, however, and began to operate in near-secrecy, the conflict escalated dramatically. It will be up to the Clinton administration, in which for the first time in twelve years one party controls both branches, to reduce conflict to more tolerable levels.

The Clinton administration took a step toward putting its stamp on regulatory policy with an executive order on regulatory planning and review, issued in September 1993. The order retained the requirement for cost-benefit analysis of major rules and OMB's authority to review proposed and final rules before agencies issue them. In these two key provisions, the Clinton order continued the policies of the Reagan and Bush administrations. The new executive order differed, however, in other respects. It gave agencies more discretion in using qualitative data to compare costs and benefits, and it specifically listed "equity" as a criterion for evaluating regulatory policies. More important than the written provisions of the order itself was a commitment that under President Clinton OMB would be more flexible and cooperative in reviewing agency rules than it had been under the last two administrations. OMB would be more accountable for meeting time schedules in its reviews, would work with agencies to set priorities, and would give more regard to agency discretion in implementing regulatory statutes. Early signals were that agency–OMB relations would improve under President Clinton but that OMB still would take its regulatory review role seriously.

OTHER FEDERAL AGENCIES

Environmental programs are concentrated in EPA, but other agencies in the federal government affect the environment. A major trend in recent years has been to recognize the need to integrate environmental with other policy sectors—agriculture, energy, transportation, trade, and land management, among others.

Relations between EPA and other federal agencies usually fall into one of four patterns. In the first, EPA and other agencies share regulatory or research roles. In the second, agencies work jointly to respond to emerging issues, like global warming. In the third, agencies like the Department of Agriculture or the Department of the Interior manage programs that may affect environmental quality. In the fourth, such agencies as the Department of Defense or the Department of Energy

(DOE) may cause and have to deal with problems, such as hazardous waste cleanups at nuclear weapons sites, over which EPA has regulatory authority. Each of these patterns is considered briefly below.

Shared Regulatory or Research Roles The pattern here is usually one of cooperation in dealing with problems for which EPA and other agencies share regulatory authority or research needs. The Environmental Protection Agency, the Food and Drug Administration (FDA), the Occupational Safety and Health Administration (OSHA), and the Consumer Product Safety Commission (CPSC) regulate chemicals or products that pose risks through a variety of exposures. Chemicals like asbestos, vinyl chloride, and benzene are a threat in both occupational and ambient settings. In reacting to evidence of dioxin risks from paper products and manufacturing, for example, EPA and FDA agreed on a joint strategy: FDA investigated consumer risks from paper products and EPA regulated discharges from pulp and paper mills.

Emerging or Rapidly Changing Issues As new issues emerge, agencies find it necessary to integrate policies more closely than in the past. An example is the issue of global warming. Concern over the effects of greenhouse gases grew rapidly in the late 1980s and early 1990s. Within the U.S. government, several agencies had to work together closely to increase research on warming, explore policies for stabilizing emissions, anticipate the effects of warming and evaluate strategies for adapting to them, and take part in international research and policy debates. The result is closer relations between EPA and State on diplomacy; Energy, Agriculture, and Interior on policy; and the National Oceanic and Atmospheric Administration (NOAA), Geological Survey, and Office of Science and Technology Policy on research.[26] Similarly, the emphasis given to environmental issues in trade policy—especially the debate over NAFTA—expanded cooperation between EPA and the Office of the U.S. Trade Representative.

Programs that Have Environmental Effects In many cases, agencies make policies or manage programs that have environmental effects. An example is the Interior Department. Its Bureau of Land Management (BLM) manages one-third of the nation's land, on which it oversees such activities as mining, cattle grazing, and water resource management. Also in Interior, the Bureau of Reclamation builds and operates irrigation and other water projects that affect water use, soil quality, water quality, and the survival of wetlands. The Minerals Management Service awards leases and sets environmental standards for oil and gas drilling on federal lands, including the Outer Continental Shelf.

Another agency with programs that have an impact on the environment is the Department of Agriculture.[27] Modern agriculture affects environmental quality in several ways. Intensive cultivation and limited crop rotation deplete nutrients in the soil and contribute to soil erosion. Fertilizers, pesticides, and animal wastes wash off the land with sediment that pollutes surface waters, or they percolate down to contaminate aquifers that are used as sources of drinking water. Wetlands and other habitat are lost when land is converted to production, often as marginal cropland. For several reasons, most of them related to maintaining high production and the income of farmers, U.S. policies have encouraged these effects. More recently, especially through the Food Security Act of 1990, EPA and Agriculture have taken steps to make farming more benign environmentally. Still, EPA's mission of reducing pollution and conserving natural resources often conflicts with USDA's mission of promoting agricultural production.

Problems Caused by Other Agencies In other cases, agencies are the cause of environmental problems. Their relationship to EPA may resemble that of industry more than that of another agency, although their political status is very different. A controversial and expensive example is cleanup of nuclear weapons plants that are managed by the Energy Department. At sites like Rocky Flats (Colorado), Hanford (Washington), Fernald (Ohio), and Savannah River (South Carolina), multibillion-dollar cleanups affecting health and ecological quality lie ahead—with DOE the responsible party. The National Commission on Environmental Quality goes so far as to describe the Departments of Energy and Defense as "the largest polluters in the United States."[28] Costs will be huge, perhaps amounting to $150 billion over the next thirty years.[29]

Because of the doctrine of sovereign immunity, which shields government entities from lawsuits, EPA had limited authority over these sites for some time. Congress expanded EPA's authority over these cleanups in the Federal Facilities Compliance Act of 1992, however, and Energy and Defense have shown signs of taking their cleanup obligations more seriously in the Clinton administration.

THE COURTS AND JUDICIAL REVIEW

The courts play a role in policy making in this country that greatly exceeds that of their counterparts in other industrial democracies. Patterns of policy making, the legal culture, rules of standing and review, a spirit of activism from many judges and courts—all of these factors combine to produce a judiciary that is a lively player in decisions and a major force in policy.[30]

THE CONTEXT FOR JUDICIAL REVIEW

The institutional perspective of the federal courts differs from that of the president or Congress. Consider some differences. Federal judges are not elected; they are appointed by the president and confirmed by the Senate. They enjoy lifetime appointments and are insulated from many day-to-day pressures that are faced by the executive and legislative branches. Because they sit on courts of general jurisdiction, federal judges do not specialize; they hear a criminal case one day, an environmental one the next, a complex securities litigation on another. They are powerful, and they can greatly influence agency policies and priorities, but they are constrained by elaborate procedural rules and subject to oversight themselves through the appeals process. They decide issues in the context of a highly structured, adversarial process, and they can act only when litigants bring cases to them for decisions.[31]

There are three levels of federal courts, with the Supreme Court at the apex. Environmental issues rarely reach this level, but when they do, the Supreme Court can have sweeping effects on policy. Below the high court are courts of appeals for the thirteen federal circuits. Under most agencies' enabling statutes, challenges to regulatory decisions by federal agencies are usually filed in one of these courts of appeals. Federal rules of jurisdiction allow parties to file either in the circuit in which they are located or in the Circuit Court of Appeals for the District of Columbia, because it is the seat of the federal government. Below the courts of appeals are the federal district courts, which have original rather than appellate jurisdiction. Of all the courts, the D.C. Circuit is the most influential on environmental issues—and on administrative and regulatory matters generally.

Judicial review and administrative law have gone through several phases in the history of U.S. regulatory agencies.[32] In pre–New Deal days, the focus of judicial review was on protecting the rights of private parties against intrusion by administrative agencies. Judges viewed agencies as impartial "transmission belts" that translated the intent of Congress into administrative actions. The courts evaluated agencies on the basis of the fidelity of their actions to the intentions of Congress. This view began to break down during the New Deal explosion of agencies and statutes in the 1930s, when Congress delegated broad authority directing agencies to act in the "public interest." Given these open-ended grants of authority, it was difficult for the courts to evaluate an agency's actions according to its fidelity to Congress's intent. In place of a transmission belt model, judges relied on a model based more on agency expertise, and they often deferred to that expertise in their reviews.

This expertise model too began to break down, in the 1960s, as judges saw that agencies' actions often involved more than the impartial application of expert knowledge. Courts showed concern about the rights or interests of groups that were not represented in agency proceedings. Judges turned from an expertise to an interest representation model. The test of an agency's decision processes was "fair representation for all affected interests in the exercise of the legislated power delegated to agencies."[33] The courts looked more at process and less at substance. Judges developed what was known as the "hard look" rule, requiring agencies to "consider all relevant factors, to respond to criticisms, to evaluate the alternatives, and to explain the manner in which they had done so."[34]

By the 1980s, in part because of a trend toward deregulation, the courts began to move back to the fidelity doctrine (evaluating agency decisions on the basis of fidelity to congressional intent) that had characterized their review decades before. Agencies had discretion but must exercise it reasonably, based on a reading of congressional intent. Courts still looked at the process by which agencies made decisions, but they also looked closely at substance. Agency action is justified, in this view, when "it remains faithful to the dictates of the legislative process."[35] As a result, agency decisions are more likely to be reversed and returned to the agency for reconsideration. Table 3 gives an overview of these four phases in judicial review of administrative actions.

EFFECTS OF THE COURTS ON AGENCY POLICY MAKING

Even a quick review of the record confirms that the courts "have had a significant effect on the policies and administration of the EPA."[36] In an analysis of two thousand EPA cases decided between 1970 and

TABLE 3 MODELS OF JUDICIAL REVIEW OF ADMINISTRATIVE ACTIONS (1920s–1980s)

Model of Judicial Review	*Time Period*	*Judicial Role Was to*
Transmission belt model	pre–New Deal	Ensure the "fidelity" of administrative decisions to congressional intent Protect the rights of private parties against agency intrusion
Agency expertise model	1930s and beyond	Acknowledge the expertise of agencies Defer to agency expertise, except when private rights or clear congressional intent threatened
Interest representation model	1960s and 1970s (but it remains influential)	Ensure fair representations of all parties affected by agency decisions Take a "hard look" to see that agencies considered and explained all relevant factors
Revived fidelity model	1980s and beyond	Look again at fidelity to congressional intent, especially in "deregulatory" actions Ensure that processes are fair *and* that agencies exercise their discretion reasonably

SOURCE: Adapted from the discussion in Merrick B. Garland, "Deregulation and Judicial Review," *Harvard Law Review* 98 (1985): 505–591. Copyright © 1985 by the Harvard Law Review Association.

1988, Rosemary O'Leary found several patterns in how court decisions have influenced EPA. In one pattern, merely filing a case provoked a response from EPA by getting its attention on an issue where it was vulnerable to legal challenge. An example is the withdrawal of an EPA rule in 1982 that allowed the disposal of liquid hazardous wastes in landfills. In other patterns, a court fully upheld the agency (causing no change in policy) or made a straightforward decision that changed policy. An example of the latter is a 1981 decision that accelerated issuance of the first full set of the RCRA standards.[37] In a third pattern, there was more active use of judicial discretion. In some of these cases, a court used its discretionary powers to push EPA to take an action that it might not have taken on its own. In other cases, O'Leary observes, judges have gone further and "have overstepped their statutory bounds, have questioned the scientific and policy expertise of EPA staff, and have aggressively pushed the EPA into action."[38]

Of the many ways in which courts have significantly affected EPA, three stand out. The first is by setting or reshaping agency priorities. Often, "compliance with court orders has become the Agency's top priority," with courts deciding what issues will get attention.[39] An example is a 1984 decision ordering EPA to set effluent guidelines (technology-based discharge standards) for toxic pollutants under the Clean Water Act.[40] This ruling committed EPA to a ten-year effort to establish effluent guidelines for many categories of dischargers.

Another way the courts influence policy is by redefining the relations between EPA and other agencies. In 1984, for example, a federal district court concluded that the CWA and the RCRA applied to operations at the Department of Energy's facilities; as a result, the court ordered DOE to apply for EPA permits. Another court ruling changed the delicate relations between EPA and OMB under E.O. 12291 when the judge held that OMB review could not legally cause EPA to miss a court-ordered deadline for issuing a rule. Since the ruling in EDF v. Thomas, OMB has been more careful about holding up rules.[41] Court deadlines give EPA and other agencies leverage in negotiations with OMB.

A third way the courts affect agencies is by defining the analytical basis for agency policies. A court remand of an EPA rule banning most uses of asbestos in the United States is an example.[42] In this case, the U.S. Court of Appeals for the Fifth Circuit found that in issuing the asbestos ban, EPA had failed to meet the TSCA's "unreasonable risk" standard. One of the court's criticisms was that EPA had not sufficiently evaluated the risks posed by possible substitutes for asbestos. The court also required EPA to base its regulation almost entirely on evidence of the *quantitative* risks of asbestos; it did not give weight to EPA's arguments regarding the ban's *qualitative* benefits. In addition, the court concluded that EPA had failed to show that a near-total ban on asbestos was the least burdensome way to protect the public against unreasonable risk. By requiring quantitative evidence of risk, a full analysis of the risks of asbestos substitutes, and a more complete showing that alternatives to a ban would have been ineffective in reducing asbestos risks to "reasonable" levels, this ruling greatly increased the burdens of justifying regulation under the TSCA, even when evidence on risk is strong, as it was with asbestos.[43]

Within an agency like EPA, the specter of judicial review increases the power of the legal staff. Indeed, the anticipatory effects of the courts may be as important as the direct effects of their decisions. Internal policy deliberations constantly take the likely reactions of the courts into account. The general counsel's office takes part in all internal regulatory work groups and often drafts rules and support documents. The general counsel usually is a close adviser to the administrator. Anticipations of judicial reactions pervade the policy process.[44] In addition, anticipated reactions from the courts and the need to establish a strong record increase an agency's efforts to justify policies with elaborate record keeping and legal justifications. "A vicious cycle soon develops: the more effort an agency puts into building a record, the more it fears that effort going to waste, and the more effort it makes to cover all possible bases."[45] R. Shep Melnick argues that judicial review reduces the power of political executives within agencies. They "are less able to set priorities or to resist demands from congressional committees and interest groups."[46]

It would be difficult to characterize the effects of the courts as entirely positive or negative. Their most positive effect has been to impose a standard of procedural rationality on decision making by EPA and other agencies. Judicial review makes agencies follow predictable procedures, take into account a range of arguments and evidence, justify decisions with good information, and keep agency policies reasonably consistent with enabling laws. And yet the effects of court review are not always benign, as we have seen. Judicial decisions can skew agency priorities, though a prudent use of judicial discretion merely applies those priorities or directions Congress already has given. And just as they serve to lend some procedural validity to agency policies, the courts may also lead agencies to be unduly cautious in their desire to avoid reversal. In the deregulatory politics of the 1980s, the courts added weight to the efforts of public interest groups and Congress in a struggle with the White House to control environmental policy.

INTERGOVERNMENTAL PARTICIPANTS IN POLICY MAKING

Environmental policy is made in a complex intergovernmental setting. My subject here is national policy, but what happens in Washington is matched in importance by what goes on in Albany or Austin—and in capitals around the world. This section looks at state–federal relations in environmental management and at the institutions that make international environmental policy.

STATE–FEDERAL RELATIONS

In France, educational administration is highly centralized. It used to be said that education in that country was so tightly controlled that at any moment each child in every school in the nation was studying the same subject. Contrast this with American education, where curricula are set locally or at the state level and there is a strong tradition of local control.

Although educational policy is decentralized in the United States, environmental policy is much less so. As the national environmental programs expanded in the late 1960s and the 1970s, what had been left to the states or to the old pattern of cooperative federalism became subject to more national control. State control was replaced by "conjoint federalism" and national regulation. Of twenty-five federal environmental laws passed between 1960 and 1980, David Welborn notes, eighteen asserted federal authority on matters in which states had previously held sway.[47]

State–Federal Authority Federalism—the sharing of power among levels of government—is a significant feature of policy making in the United States. Over the years, scholars have used many terms to describe the paths that federal–state relations may take, from "marble cake" to "picket fence" to "cooperative" federalism. In environmental policy, Welborn's concept of conjoint federalism describes the state–federal relationship most aptly.

Conjoint federalism describes a relationship in which state and federal authority blend and apply concurrently to the objects of regulation. National goals drive regulatory programs. National policy forces state action. Federal and state agencies exercise concurrent enforcement authority. If a state demonstrates that it can meet national program requirements, EPA grants it authority to implement programs at the state level. However, this authority is subject to close federal oversight. EPA can revoke the state's authority if it judges the state's program to be inadequate. If a state refuses to participate, or is unable to meet the requirements for delegation, EPA takes control of the program.

Currently, with the exception of highly centralized programs like mobile source air regulation and pesticides registration, most of the national programs fit this pattern of conjoint federalism.[48] Under the Clean Air Act, for example, EPA sets national standards for toxic air emissions, then delegates authority to the states to apply them (if states meet the standards for delegation). States prepare State Implementation Plans showing how they will regulate existing sources to meet ambient air goals, which are set in Washington. EPA reviews the SIP (and any subsequent changes a state proposes to make) and approves, disapproves, or requests changes in the plan. A state's failure to adopt a plan or implement an approved one may cause EPA to issue a federal plan in place of the SIP, as EPA proposed to do for the Los Angeles portion of the California SIP in 1990.[49] The pattern for the water and hazardous waste programs is roughly the same.[50]

So in some areas—air, surface water, and hazardous waste—the framework and often the specifics of programs are defined by EPA. EPA sets the minimums for state programs, must approve their design and implementation, and may impose sanctions if states fail to meet the national program requirements. In other areas—municipal waste (garbage), groundwater, and nonpoint source water pollution—states have more discretion. Yet even this has changed. In 1991, EPA issued standards for the design and management of municipal waste landfills, as it was directed to do under the 1984 amendments to the RCRA. Similarly, the Water Quality Act (in 1987) set out some required elements for state nonpoint source water programs.

Table 4 presents my own view of the relative degrees of centralization of several programs. At the left are the programs that are highly centralized at the federal level: mobile source air pollution, pesticides registration, the approval of new chemicals. Toward the middle, there is a mix of federal and state roles—the pattern of conjoint federalism discussed above. The programs for regulating hazardous waste, major new sources of air pollution, and industrial point sources of water pollution fall into this range. Other programs are relatively decentralized, and the states exercise more control. The nonpoint source water, groundwater, and municipal solid waste programs fall more toward this end of the continuum. Of course, there are exceptions. California, for example, because of the severe air quality problems in parts of the state, sets stricter limits on auto and truck emissions than the federal government does.

The most serious effort to change this pattern of conjoint federalism was President Reagan's "New Federalism" initiative in the 1980s. Its themes were decentralization and defunding.[51] Based on the premise that the federal government had assumed too large a role and that states could implement many programs more effectively and responsively, the New Federalism aimed to return authority to the state level. At the same time, it led to major cuts in grants-in-aid to help the states manage programs. In the environmental area, the result was to accelerate the delegation of authority to the states and to reduce federal oversight. Large cuts in aid had a major effect on the states, many of which had depended on grants to fund large parts of their programs.[52]

TABLE 4 VIEW OF THE RELATIVE DEGREES OF CENTRALIZATION IN SEVERAL ENVIRONMENTAL PROGRAMS

More Centralized	*Mixed (Conjoint Federalism)*	*More Decentralized*
Mobile Source (except California)	Hazardous Waste	Nonpoint Source Water
Pesticides Registration	New Industrial Air	Groundwater
New Toxic Substances	Industrial Point Water	Municipal Waste
	Existing Industrial Air	

NOTE: This table lists programs where there is a federal presence. Many environmental activities regarding land use, noise control, and waste management, among others, are handled entirely at state and local levels.

Although many states have increased their own capacities for environmental management as a result of the Reagan years, the old pattern of conjoint federalism still describes the state–federal relationship. The New Federalism disrupted old patterns but did not fundamentally change them. Neither the Water Quality Act of 1987 nor the 1990 CAAA departed from the past, and in some ways—for example, the treatment of water pollution from nonpoint sources—they even expanded the federal government's role.

Capacities of State Programs States vary in their commitment to environmental goals and the quality of their programs.[53] California, New York, New Jersey, Minnesota, and Wisconsin are known to have strong environmental programs that are run by capable and effective agencies. Environmental agencies in other states suffer from limited political support, inadequate funding, or poor institutional capacities. We can compare state programs in several ways. One is by looking at spending. On a per capita basis (for the period 1969–1980), California, Delaware, Maryland, Ohio, and Vermont spent the most on their programs. Texas, South Carolina, Mississippi, Oklahoma, and Utah spent the least.

Spending is only one, admittedly narrow criterion. Another way to compare states is by combining spending with other factors into a set of indicators (23 of them in the study cited here) that allow us to assess the strength of state policies for land use and the environment. Again, California, Oregon, Minnesota, and New Jersey ranked high; Idaho, New Mexico, and Mississippi ranked low. Certainly these rankings are subject to debate as many states have made serious efforts to upgrade their programs. The point is that there is variation from state to state in the quality of their programs and the levels of fiscal and political support for them.[54]

A pressing issue in state–federal relations is the states' financial capacities for complying with federal mandates (these are regulatory requirements that states must implement or be subject to federal sanctions). As the costs of environmental programs have grown, the federal share of program costs has shrunk. Between 1979 and 1989, federal air, water, and waste grants to states fell almost 30 percent in real terms.[55] States have had to pick up a larger share of the funding for programs that previously were covered by federal grants. This is partly due to the loss of funds from the New Federalism. But it is also due to several new regulatory mandates that the federal government has imposed on the states, especially in the areas of drinking water treatment, wastewater treatment, municipal waste, and sewage sludge disposal. To respond to these mandates, states are turning to new financing mechanisms—fees, environmental taxes, bond issues, and revolving loan funds—to supplement what they can finance out of state general revenues. Many states have also called for increased federal funding and more flexibility in setting their own environmental priorities, so they are not forced in all cases to follow the federal mandates.[56]

The capacities of state programs and the division of power between federal and state levels remain major issues. State capacities have improved greatly in the last two decades. There was a time when the leadership and innovation in environmental programs came mostly from Washington, but now states like California, New York, New Jersey, and Illinois are leaders in their own right. The selection of Carol Browner to lead EPA under President Clinton guarantees that state–federal issues will remain high on the agenda.

INTERNATIONAL INSTITUTIONS

A clear lesson to emerge from the 1980s was the irrevocably international character of most environmental issues. Acid rain, ocean pollution, deforestation, stratospheric ozone depletion, trade in pesticides and toxic wastes, and global warming are only some of the problems whose causes and solutions extend beyond national boundaries. International and regional organizations are part of the institutional landscape. This section describes two kinds of international institutions, the United Nations Environment Programme (UNEP) and the multilateral development banks, to set the stage for a discussion of international environmental policy.

United Nations Environment Programme UNEP has emerged as a focus for organizing international efforts on the environment.[57] It was formed by the General Assembly in 1972 as an outgrowth of the UN Conference on the Human Environment, held earlier that year in Stockholm. Led by an executive director and based in Nairobi, UNEP is a "nonexecuting" agency within the UN system, which means that it has limited staff and authority. UNEP is separate from the UN's specialized bodies; the latter were created by treaties among member governments rather than by the General Assembly itself. Yet much of UNEP's influence derives from its ties to such bodies—the Food and Agriculture Organization (FAO), the World Meteorological Organization (WMO), the World Health Organization (WHO), and the World Bank.

UNEP serves the international community as a facilitator, catalyst, and clearinghouse for research and information. An example of its information role is the Global Environmental Monitoring System (GEMS), which it maintains with the International Council of Scientific Unions (ICSU). But UNEP's most visible achievements have been in promoting international agreements. Among these are the Convention and International Trade in Endangered Species, the Convention on Prevention of Marine Pollution (London Dumping Convention), three conferences on the Law of the Sea, the Conventions and Protocols of the Regional Seas Programme, and the Vienna Conventions and Montreal Protocol on Stratospheric Ozone. UNEP has also served as the focal point for international responses to the evidence of global warming. With WMO, UNEP was instrumental in forming the International Panel on Climate Change (IPCC) in 1987. IPCC set up working groups to evaluate scientific information, examine socioeconomic and environmental effects, and formulate strategies for reducing greenhouse gas emissions and adapting to climate change.[58] Most recently, UNEP was the organizing body for the UN Conference on the Environment and Development (UNCED) in Rio de Janeiro, which is discussed below.

Multilateral Development Banks For decades, the multilateral lending banks have been a major force in international development. The largest and most important is the World Bank (founded in 1946).

Along with the International Monetary Fund and the Inter-American (1959), the African (1964), and the Asian (1966) Development Banks, the World Bank accounts for tens of billions of dollars in aid to developing nations each year. Many people see the banks and their policies both as a threat and as a partial solution to environmental problems in the developing world.[59]

Multilaterals affect the environment in developing nations in two ways. First, many bank loans finance environmental projects in developing countries—projects for reforestation, sewage treatment, safe drinking water, or restoration of wetlands, to name a few. The amount of these loans has grown in recent years, in response to criticism that the banks spent too little on the environment.

Second, many bank-funded projects designed to promote economic growth (the banks' primary goal) have caused environmental damage. The banks have usually funded large-scale, capital-intensive projects: highway construction, hydroelectric and irrigation projects, or massive resettlement programs to stimulate growth in undeveloped areas, such as the interior of Brazil. For example, because they emphasize high yields and the production of cash crops (like tobacco) for export, bank-financed agricultural projects favor high-input strategies, leading to the heavy use of pesticides and fertilizers and intensive cultivation of croplands. The results of such strategies include the loss of soil nutrients, soil compaction, inundation of forests, salinization of irrigated lands, and the loss of tropical forests at a rapid rate.

The banks have responded to criticism that they have funded too few environmental projects and that their development projects have caused environmental damage in several ways. The World Bank, as an example, now requires more evaluation of the environmental impacts of its projects when it makes decisions about loans. To better assess environmental effects, the bank appointed an environmental adviser in 1970 and created an environmental affairs office in 1973 and a central environmental department in 1987.[60] The regional banks have taken similar steps. Observers agree that the banks now are more sensitive to the environmental effects of their loan decisions than they have been in the past.

But critics argue that procedural and organizational change is not enough. They define the problem in more structural terms—as too much reliance on a Western model of economic development and too much dependence on the structure of the international economy. In this view, the banks will need to make more radical reforms that stress the sustainability of the resource base over the more conventional, Western-style model of economic growth.[61]

UNCED in Rio Easily the most important international environmental event of the last decade was the United Nations Conference on the Environment and Development, held in Rio de Janeiro, Brazil, in June 1992. Organized by UNEP and attended by representatives of some 180 countries, UNCED marked the twentieth anniversary of the Stockholm meeting. Although a direct descendant of Stockholm, UNCED differed from its parent in several ways. Stockholm recognized and reinforced environmental movements that were emerging in the late 1960s and early 1970s. Rio came at a time when these movements had matured and environmental problem solving had moved to a global stage. Only two prime ministers were at Stockholm; Rio was an environmental summit. With over one hundred heads of state attending, including President Bush, Rio was the largest gathering of world leaders ever. Known also as the Earth Summit, Rio drew a lively array of interest groups, activists, and publicity seekers as well as the delegates.[62]

Hyperbole was in the air at Rio. The National Audubon Society called it "the most important meeting in the history of mankind." UNCED's chief organizer, Maurice Strong of Canada, saw the stakes as "nothing less than the fate of the planet."[63] Yet tangible issues were on the table. Five documents came out of the summit: a climate treaty, in which countries agreed to meet voluntary targets for reducing greenhouse gas emissions; a biodiversity treaty to protect plants, animals, and habitat threatened by extinction from deforestation or pollution; the Rio Declaration on Environment and Development, which set out global intentions to achieve an environmentally sustainable world economy; a declaration of support for forests; and "Agenda 21," an inventory of problems combined with an action plan for the next century.

It was the first two treaties—on climate and biodiversity—that were the most important substantively. The climate treaty committed signatory nations to reduce greenhouse gas emissions to their 1980s levels by 2000. Yet the United States refused to go beyond making a voluntary commitment, to the disappointment of Japan and the Europeans, giving the treaty less force than it otherwise may have had. In final negotiations in New York, which were concluded just before the Rio summit opened, the United States convinced 142 other nations to agree to a treaty without firm commitments, as a condition of U.S.

approval. It was, *The Wall Street Journal* reported, "a vaguely worded pact that sets no binding timetables for reducing emissions, makes no commitment to achieving specific levels of emissions—indeed, makes no commitments to do anything at all."[64] Even worse for nations expecting more U.S. leadership, President Bush refused to sign the biodiversity treaty. He was concerned that it would infringe the patent rights of U.S. companies. This decision drew substantial criticism, as the United States was the only country at the Rio conference that did not sign.

The problem for the United States (and for EPA Administrator Reilly, as head of the U.S. delegation) was that domestic political concerns drove the Rio agenda. The administration had already decided to press economics over the environment in the reelection campaign, as the regulatory moratorium showed, and felt a need to shore up its conservative constituency after a difficult primary season. By not agreeing to more than voluntary emission reductions in the climate treaty and by stressing the commercial interests of American firms above species preservation in the biodiversity treaty, the Bush administration was responding to voter concern about the economy, which it saw as the main issue in the fall election.

Environmental policy has become as important as economic, trade, or national security policy in the international arena. In the 1980s, for example, the United States put eleven multilateral environmental treaties into force, as many as had entered into force in the previous three decades combined.[65] Issues like climate change, losses of tropical forests and rare species, the effects on trade of environmental programs, and the interdependence of national economies will keep environmental policy high on the international agenda. A stronger UNEP, an evolving role for the multilateral banks, greater emphasis on environmental issues in negotiations, and growing influence by environmental ministers in their own cabinets—all are likely trends. The "traditional grays of diplomacy will acquire more of a green cast" in years to come.[66]

NONGOVERNMENTAL INFLUENCES ON POLICY

Administrative policy making in the United States takes place in what organization theorists call an open system. Pluralist politics allows many avenues for access to people and groups outside the formal government apparatus. Environmental groups, industry or trade groups, and the media influence policy in important ways.

ENVIRONMENTAL GROUPS

National, nonprofit groups dedicated to environmental conservation date back to the early part of the century. The oldest and one of the most powerful is the Sierra Club, founded in the 1890s by the naturalist John Muir. The modern environmental movement dates to the 1960s and 1970s. Founded in 1970, the Natural Resource Defense Council (NRDC) was a prototype of the new environmental organization. With groups like the Environmental Defense Fund (EDF) and Friends of the Earth (FOE), it ushered in a new era of environmental activism.

Although these groups function largely as advocacy and political action groups, their success depends on maintaining a degree of scientific credibility, and many have very capable technical people as well as lawyers and lobbyists on their staffs. People from these groups often serve on technical advisory bodies for EPA and other agencies. Their advice and analysis are sought by Congress, especially by members with an environmental agenda. Leading environmentalists sometimes serve inside government: the AA for air under President Carter had been a senior NRDC official; the AA for policy under President Clinton had most recently been the legislative director for the Sierra Club before coming to EPA.

Lobbying Congress is one tool environmental groups have used effectively. Another is litigation. Groups like NRDC have skillfully used the courts to influence policy. The National Environmental Policy Act enabled outside groups to challenge the adequacy of Environmental Impact Statements (EISs) in court. Major changes in administrative law in the 1960s and early 1970s opened up access to the courts for environmental groups.[67] A series of court rulings expanded the rules of standing that allow public interest groups to sue agencies. The major environmental laws, most passed in the 1970s, included broad provisions for public participation. The plaintiffs in court cases often are environmental groups, who use judicial decisions to change policy and draw attention to issues that are important to them.

In addition to litigation and lobbying, environmental groups can use the media to affect agency policies. An example is the campaign that NRDC mounted in 1989 to have EPA cancel registration of the pesticide Alar, which was used on apples to extend their storage life. Earlier studies had shown possible cancer risks from residues of Alar on apples, but EPA had never felt that the evidence was sufficient to cancel its use. NRDC commissioned an analysis of the health risks of Alar and found somewhat higher risks than EPA had found. What was most effective was not the NRDC analysis itself but how it was publicized. As *The Washington Post* observed, NRDC "achieved more in a few days than all of the lobbying and litigating since Alar was found to cause cancer in laboratory animals in 1985."[68] Stores pulled apple juice from shelves, schools dropped apple products from their menus, and the manufacturer took Alar off the market.

INDUSTRY INTERESTS AND TRADE ASSOCIATIONS

For at least the first decade of the national programs, relations between industry and government can best be described as hostile. Most regulation was directed at the major industrial sectors—steel, automobiles, utilities, chemicals, and others. Environmental goals were seen to conflict with economic ones. The debate often took on moral overtones. To environmentalists, industry was irresponsibly poisoning the air and water in pursuit of selfish economic interests. To industry, environmental groups had set out to debilitate the economy with excessive regulation and constraints on growth. The energy crisis, high inflation, and foreign competition reinforced the perceived conflicts between economic and environmental values through the 1970s.

Environmental and economic goals may still conflict, but the atmosphere has changed. Regulation is more accepted today than it was in the 1970s. Firms strive to show how responsible they are: the 3M Corporation is a leader in integrating pollution prevention into its activities; firms like McDonald's get public relations value out of suspending use of chlorofluorocarbon (CFC) products or reducing the volume of packaging; Amoco worked with EPA on a joint analysis of cross-media planning at a refinery.[69] Among the reasons for this trend are a recognition that pollution programs enjoy public support and are here to stay; an awareness of the public relations value of sound environmental management; and an appreciation of the economic benefits of pollution prevention and control, such as less fuel use, more recovery of materials, and lower insurance and liability costs.

This is not to say that business groups do not challenge government policies regularly. There are differences on issues like the scope of regulation, the stringency of controls, the flexibility allowed for compliance, or the costs in any program. At times, when proposed actions may threaten the survival of an industry, the opposition can be intense. In issuing its asbestos phase-down and ban rule, for example, EPA faced intense lobbying from the asbestos industry and the government of Canada, where most asbestos used in the United States is mined. The industry challenged the ban in court and succeeded in having it overturned.

Industry participates in policy making in several ways. Large corporations like General Motors, DuPont, Shell Oil, and Monsanto maintain large governmental relations staffs, conduct research and analysis, and lobby and litigate. Trade associations enable firms to pool their resources. For example, the Chemical Manufacturer's Association represents the interests of the large chemical companies, the National Agricultural Chemicals Association represents pesticide producers, the National Gasoline Marketing Association represents petroleum distributors, and so on.

Another set of trade associations has become more visible and influential. As pollution control rules grow, so does the market for environmental products and services. With an annual bill of over $100 billion, environmental protection is a big business in its own right. Unlike most other businesses, firms in pollution control or environmental services have an interest in maintaining or expanding EPA's programs. Regulation increases the demand for these companies' products and services, giving them an interest in strong programs and making them allies of the environmentalists.

THE MEDIA

Media coverage increases in proportion to the salience and visibility of an environmental issue. At the local level, stories about waste sites, incinerator emissions, contaminated aquifers, or chemical plant emissions as reported in the Toxic Release Inventory draw attention to the environment. Nationally, papers like *The New York Times, The Washington Post,* and the *Los Angeles Times* shape the policy agenda and pressure policy makers to act in certain ways.

Agency heads understand the power of the media, and they sometimes use that power to their own purposes. One of the more revealing such examples is a memorandum Administrator Reilly sent to EPA staff after his return from the Rio summit. Reflecting his frustration with the failures of U.S. leadership, especially the flap over the biodiversity treaty, and perhaps his own policy differences with the administration over the previous year, he wrote about the Rio summit, "For me personally, it was like a bungee jump. You dive into space secured by a line on your leg and trust it pulls you up before you smash to the ground. It doesn't typically occur to you that someone might cut your line!" Though Reilly's staff claimed that these observations were meant for EPA staff only, only the rawest political novice could think that a document sent to 18,000 people would not find its way into the press. It had to be a deliberate act to get Reilly's views—and his frustrations—on the public record, and it received a fair amount of press coverage in the aftermath of the summit.[70]

CITIZEN PARTICIPATION AS AN INSTITUTIONAL CHALLENGE

One especially important aspect of the institutional setting for environmental policy is public participation in environmental decisions. In most respects, access to policy making is greater in the United States than in the other industrial democracies. Broad judicial rules of standing and a tradition of active court review offer ways for outside groups to challenge agencies and block agency actions. Opportunities are also provided by active legislative oversight and provisions for public hearings and other forms of consultation.

Yet both litigation and public hearings have disadvantages as mechanisms for participation. Hearings are loosely structured, open forums, where interested members of the public hear agency proposals and respond, typically in a format that is set by the agency. Hearings often are held late in a decision process, after options have been narrowed and the important choices have been made. Participation is usually strongest from the opponents of a policy, and even then from well-organized groups with greater resources. Hearings have come to be equated with opposition, which causes many agency officials to look on them with disdain and citizen groups to see them as a last stage for vetoing proposals and preparing for lawsuits. So hearings have a place, but on their own they are a limited solution to the need for participation.

What are the alternatives? Other mechanisms for citizen participation include public surveys, voter initiatives, negotiated rule making, citizen review panels, advisory commissions, written comment processes, and site-specific dispute mediation. All have their strengths and weaknesses. Initiatives, for example, place issues on the ballot for voter approval. California often makes major decisions through initiatives. An example is Proposition 65, passed in 1986, making warning labels and other measures mandatory for products that contain carcinogens. Initiatives give a snapshot of public opinion on issues, but they cannot reflect the intensity of opinion, and they force voters to select from among dichotomous choices (either for or against). Surveys also may convey a sense of public opinion, but they have limited educational value and fail to provide a forum for public debate or deliberation. Citizen panels and negotiated rule making offer strengths as participation mechanisms. I will return to them in the last chapter.[71]

Achieving effective citizen participation is one of the five institutional challenges I posed at the outset. People value the opportunity to influence decisions that affect them, especially in the United States, where we have a sense of "subjective competence" and expect to have a say in decisions.[72] As Francis Rourke has noted, efforts to involve citizens reflect "a deep-seated belief on the part of legislative and executive officials that bureaucratic power can best be legitimized by being democratized, by bringing the decisions of public bureaucrats much more closely under the control of private citizens."[73]

INSTITUTIONAL ANALYSIS AND ENVIRONMENTAL POLICY

The analysis of the institutional setting for environmental policy has covered many topics—from laws to EPA, from Congress to the courts, from states to international organizations and the need for citizen participation. It is a varied list of issues and players. Yet certain themes emerge.

One is that the process for making environmental policy is fragmented. The environment is not unique in this respect. The constitutional separation of powers, interest group pluralism, an assertive Congress, activist courts, federalism, and a pattern of reacting to problems as they emerge rather than comprehensively all affect most public policy issues in the United States. Add to this the particular features of environmental policy: EPA's bureaucratic origins as a holding company operating under diverse laws; oversight scattered among multiple committees and subcommittees in Congress; a mostly open door for appeals to the courts; an aggressive environmental community that skillfully uses what levers are available for shaping policy; well-organized industry interests with a lot at stake; and an internal EPA organization that both reflects and reinforces a fragmentation in constituency interests outside the agency. Policy is shaped through a series of "issue networks," which Hugh Heclo describes as "the many whose webs of influence provoke and guide the exercise of power."[74]

A second theme is institutional conflict. The struggle between Congress and the president for control over policy has been a central fact of EPA's existence for all of its two-plus decades. Although much of this conflict can be attributed to differences in party control of the executive and legislative branches, there are institutional sources of conflict that go beyond these partisan differences. Yet divided control of the government over most of EPA's existence (18 of 23 years) has clearly affected the relations among institutions. Remember, though, that even same party control of the legislative and executive branches does not eliminate conflict; the Carter administration was just as concerned about White House oversight of regulatory policy as were those that came before and after. Interbranch competition will continue—but probably not characterized as the outright conflict of the 1980s and early 1990s.

A third theme is the demands that the evolving nature of environmental problems place on institutions. In the 1960s and 1970s, the fluidity and regional dispersion of most problems made it necessary to shift much authority for making policy from local and state to national institutions. In the 1980s and 1990s, we have been seeing a similar shift in authority from national to international institutions. Environmental problems seem constantly to be running ahead of institutional capacities for dealing with them. Similarly, even within levels of government, there is a growing need to integrate policies across boundaries: environmental policy with agricultural, energy, economic, trade, transportation, and other sectors. Institutions have to be dynamic to be able to adapt to changes in problems and new approaches to solving them.

What we have seen in the last two chapters is the particular form of American environmental policy making. It can look very different in other nations and political cultures, even in other industrial democracies. In Britain and Sweden, for example, the relations between government agencies and industry are far more cooperative and based on consensus. Legislatures and courts are far less active participants, serving mainly to protect private rights rather than acting as micromanagers or partners in policy making. The public is less likely to challenge agency decisions, because of traditional deference to agency authority and technical expertise. Administrative processes in Europe are less complex, and there is not the same obsession with scientific analysis as "proof" for justifying or challenging decisions. Despite these differences, the policy outcomes are not all that dissimilar. A study of toxic chemical regulation in the United States and Europe, for example, concludes that the different cultures and policy making processes "have led to remarkably similar policy choices, particularly in the selection of specific chemicals as targets of regulation."[75] Although I will not discuss in detail cross-national aspects of policy making here, they are a fruitful area for additional reading.[76]

Legal arrangements and institutional forms determine how policy is made and implemented and with what degree of success. Protecting the environment is more than a matter of science and technology, water sampling, and air monitoring. It is a matter also of politics, laws, and relations among institutions. And it should be a matter of good analysis, as we will see next. Chapter 4 examines the two principal kinds of analyses used in making environmental decisions (risk assessment and economic analysis), their strengths and limitations, and their value in achieving the kind of bounded rationality to which many policy makers aspire.

CHAPTER FOUR

Analyses

In the discussion of models of policy making, I compared the rational and incremental models. In the rational model, goals are clear and agreed upon; policy makers have complete and reliable data; problems are well defined; a full range of policy options is identified; effects of options are understood and predictable; and final choices maximize previously stated goals. All in all, this is not a bad image of how public policy should be made. We like to think of ourselves as logical and well informed, of our government institutions as well-oiled machines making sound choices, of policy as made up of well-reasoned outcomes furthering worthy social goals.

The incremental model offers a contrasting view of policy making. Here the goals may be unclear or in conflict; information is missing or unreliable; options may be poorly defined or ignored; policy emerges piecemeal, in fits and starts; and results often are different from what was intended. An incremental model presents a less appealing but more realistic view. Studies of several kinds of policy decisions—whether in budgeting, foreign policy, or other areas—support the conclusion that policy making is less rational and more incremental than we would like it to be.

This is not to say that public policy making is irrational. Most institutions and policy makers probably aspire to be as rational as possible. But in a complex world, there simply are too many choices, reconciling too many different goals, based on too little information, made by too many people, to enable us to meet the high standards of the rational model. The best that we can achieve is Herbert Simon's concept of a bounded rationality, wherein we try to move as far from the incremental to the rational as we can.

Analysis is simply one way of extending the boundaries of rationality in public policy. When they use analysis to make decisions, policy makers try to understand the problems they are dealing with, the various constraints on their choices, how one way of responding to the problems compares to others that they might use, and the overall (for society) and specific (for groups in society) consequences of their decisions. In this chapter, I examine two kinds of analysis that play a role in environmental policy. The first is risk analysis, which estimates the harm of an activity, substance, or technology. The second is economic analysis, which calculates and predicts the costs and (sometimes) the benefits of different policy goals or decisions.

I begin with the concept of risk and the process that most government agencies use to analyze health risks. Then I turn to economic analysis and its uses in environmental policy making. The chapter concludes with a look at three of many issues that come up when we use risk and economic analysis in environmental policy.[1]

RISK ANALYSIS AND THE ENVIRONMENT

The concept of risk is central to environmental policy. Nearly any environmental problem can be seen as a matter of risk, which we can define simply as the possibility of suffering harm. Of course, risks are all around us, and they are not limited to environmental causes. Driving cars, making investments, climbing mountains, starting a small business—all of these expose us to physical, financial, psychological, or other risks.[2] My focus here, though, is risk from contamination of air, soil, and water.

There are two dimensions to this notion of risk as the possibility of suffering harm.[3] The first is the probability or likelihood of the harm; the second is the severity of the harm, its magnitude or significance. When we have a choice, most of us are inclined to avoid taking risks that pose highly probable and severe harm. Knowing that 1 in 10 of the people who try to scale Mount Everest die in the effort is enough to discourage most of us from the attempt, whatever the chance for glory. The odds are bad (1 in 10) and the harm (death) is one that most of us would regard as severe. The risk of someone in this country dying in a car accident in any year is a much lower 2 in 10,000. It is a

TABLE 5 COMPARING EVERYDAY RISKS
Activities That Increase Chance of Death
by 0.000001 (one in a million)

Activity	*Chance of*
Smoking 1.4 cigarettes	Cancer, heart disease
Spending 1 hour in a coal mine	Black lung disease
Traveling 6 minutes by canoe	Accident
Traveling 300 miles by car	Accident
Traveling 1,000 miles by jet	Accident
Eating 40 tablespoons of peanut butter	Liver cancer caused by aflatoxin B
Living 150 years within 20 miles of a nuclear power plant	Cancer caused by radiation
Eating 100 charcoal-broiled steaks	Cancer caused by benzopyrene
Living 2 months with a cigarette smoker	Cancer, heart disease

SOURCE: Richard Wilson, "Analyzing the Daily Risks of Life," *Technology Review* 81 (February 1979): 45.

risk nearly all of us have decided is worth taking. To compare some sources of everyday risk, consider table 5, which lists activities that pose a 1 in 1,000,000 increased chance of death in a year. Risk is all around us, often from unlikely sources.[4]

HEALTH VERSUS ECOLOGICAL RISKS

Environmental policy makers work in the context of two broad categories of risk. The first is the possibility of harm to human health—anything from eye irritation from air pollution to death from exposure to high levels of a toxic chemical. The object of concern is people and their well-being. We describe health risks according to several features: whether the effects are acute (immediate) or chronic (long-term), how serious they are, whether they are reversible, and the numbers and kinds of people that are affected. Society typically responds quickly to evidence of acute risks, because causes and effects usually are fairly easy to establish. Most of the debate is over chronic risks, where relationships between causes and effects are harder to establish. There is uncertainty about the sources of problems—or whether there is a problem at all.

Until recently, it was likely that when environmental policy makers referred to a chronic health risk they meant cancer. The concern about cancer has dominated government risk assessment. The state of the art for assessing cancer risks is ahead of that for other chronic risks. When agencies justify regulatory action, they usually base their case on cancer, as there is more research and data to draw on. Politically, a focus on cancer has helped environmental agencies generate public support for their programs. To ancients and moderns alike, James T. Patterson has written, cancer has been seen as "voracious, insidious, and relentless."[5] By casting itself as a cancer protection agency in the late 1970s, EPA was able to sustain public support for its programs in troubled economic times. For all of these reasons, cancer has dominated the regulatory policy agenda.

Yet many noncancer health effects should concern us as well. The causes include common pollutants, for example, lead, which is pervasive in contemporary society. Exposure to lead may impair children's physical and mental development and cause high blood pressure in white males. Another example is ozone, which forms when volatile organic compounds (VOCs) from cars and industrial sources react with nitrogen oxides in the presence of sunlight. Ozone causes short-term respiratory problems and stresses the cardiovascular system. Long-term exposure may impair lung function permanently. The list of noncancer but chronic threats to health goes on; we can expect that policy makers will give more attention to such risks in the years to come, as the methods for studying them improve and concern about them grows. Table 6 lists several types of noncancer health effects that have been linked to environmental pollution and an example of each.

Another major category of risk is harm to ecological resources. When many people think of *environmental* protection, they probably are more likely to think of ecological than health risks. Fish thriving in a clean river, a clear view of the Grand Canyon, an untainted estuary full of shellfish, a white beach with no litter in sight, a tropical rain forest with tremendous diversity in its plant and animal species—all describe ecological resources worth protecting. Of course, health and ecological risks may overlap. Contaminated fish may pose health as well as ecological risks. But other problems, like emissions from a power plant that impair visibility in a park, are largely aesthetic and ecological. The distinctions between health and ecological risks are important, even when they occur as part of the same problem. They pose different kinds of questions and present choices among diverse, often competing

TABLE 6 TYPES OF NONCANCER HEALTH RISKS THAT ARE LINKED TO ENVIRONMENTAL POLLUTION

Type of Health Risk	*Example*
Cardiovascular	Increased heart attacks
Developmental	Birth defects
Hematopoietic	Impaired heme synthesis
Immunological	Increased infections
Kidney	Dysfunction
Liver	Hepatitis A
Mutagenic	Hereditary disorders
Neurotoxic/Behavioral	Retardation
Reproductive	Increased spontaneous abortions
Respiratory	Emphysema
Other	Gastrointestinal diseases

SOURCE: John J. Cohrssen and Vincent T. Covello, *Risk Analysis: A Guide to Principles and Methods for Analyzing Health and Environmental Risks* (Washington, D.C.: Council on Environmental Quality, 1989).

values. They also require different methods for estimating risks and evaluating benefits.

The differences between health and ecological risk assessment result mainly from the greater variety in the forms of life affected and the many endpoints (the range of bad things that may happen) for ecological risks. When assessing health risks, we analyze the effects on human health. For ecological risks, we look at a variety of receptors—birds, plants, ecosystems, and so on. In addition, the endpoints are more diverse in the case of ecological risks and may include the effects on organisms, populations, habitat, natural systems, and others. Ecologists account for levels of ecological organization. They sort living systems into organisms, single-species populations, multispecies communities, and ecosystems. Risks are assessed at each level.

This discussion focuses on health risks; that is where risk assessment has been used the most to make environmental policy. But interest in understanding and assessing ecological risks has been growing. Increasingly, policy makers recognize what EPA's Science Advisory Board has described as "the vital links between human life and natural ecosystems."[6] It often is necessary to distinguish health from ecological risks for analytical purposes, but the connections between the two are strong. The next section looks at perceptions of risk. Following that is a closer look at the process of assessing health risks in environmental policy.

HOW DO PEOPLE PERCEIVE RISKS?

The study of attitudes toward risks, their acceptability, and people's behavior in response to what they think is harmful is the field of risk perception. Much of the research on risk perception focuses on psychological factors. Some of it also examines the sociological and cultural influences on risk perception, which are especially important from the perspective of public policy. These cultural factors help to explain many of the differences between the lay public's intuitive evaluations and the formal evaluations of risk by experts.[7] We will look first at individual perception of risk, then at social and cultural influences on risk perception.

As for individual perceptions of risks, experimental psychologists have compared statistical (based on quantitative studies) to perceived risks. In one study, people were asked to estimate the frequency of deaths that resulted from forty-one causes, among them, disease, natural disasters, accidents, homicides, and recreation (like mountain climbing).[8] The answers revealed differences between what people *thought* was risky and what actually *was* risky. People overestimated the risks of death from unusual, catastrophic, or lesser-known sources (such as nuclear power plants); they underestimated the risks of death from common, better-known, or discrete causes (such as driving). The public's negative views toward nuclear power are shaped by the tendency to attribute high risk to lesser-known problems that could have catastrophic effects. Similarly, rare causes of death attract more attention. Botulism, for example, accounts for about five deaths a year in the United States, but the respondents in one survey thought it caused five hundred. Because deaths from botulism are reported in the media, people are more aware of them and tend to exaggerate their occurrence.

People's perceptions of risks affect their views about the acceptability of different kinds of risk. Consider the differences between perceptions of voluntary and involuntary sources of risk. The risk perception research shows that people are more willing to tolerate risks they assume voluntarily than risks that are imposed on them by others without their consent. This explains why some people might oppose a decision to site a waste incinerator near their homes yet not be concerned with the greater statistical risks of smoking or not using seat belts. They smoke or do not use seat belts by choice.[9]

People are also far more concerned about risks that are unknown, dreaded, or seen as catastrophic than risks that are better known or discrete (occurring in a large number of small events rather than in one major event). The public's intuition about catastrophic events may have a solid foundation. An example is a catastrophic accident in one community, whose effects led to what has become known as the "Buffalo Creek Syndrome." The collapse of a slag waste dam in Buffalo Creek, West Virginia, some years ago left 120 people dead and 4,000 homeless. Nearly two years later, when psychiatric evaluators studied the survivors, they found evidence of "disabling character changes" and the sense of a "loss of communality" among them, including a loss of direction and energy.[10] Studies showed similar effects after the Three-Mile Island accident, even though there was no apparent physical damage. So attitudes about risk may reflect concerns about social stability and cohesion, about effects on communities, not just individuals.

The lesson of much of the research is that people do not react to risk in a state of social or cultural isolation. Shared values and a sense of community come into play. For example, people's views about the acceptability of various kinds of risks reflect their judgments about the institutions that manage risks in society.[11] If people doubt the trustworthiness of a corporation that is proposing to build a waste incinerator, or the objectivity of the government agency that will issue an operating permit, they will be skeptical about any evidence that they face negligible risks when the incinerator starts operating. People also evaluate risks on the basis of perceived fairness in their distribution. If a local government decides to build a landfill in a poor community and if the waste was produced mostly by a more affluent community in another part of town, we can expect more opposition based on the clear inequities of the decision. If a community views the result of a decision and the process for making it as fair, it is more likely to accept the result.[12]

Why does all of this matter in a discussion of analyses? Because there is a gap between the products of the experts in quantitative risk assessment and the informal, more intuitive evaluations of risk by the lay public. As the next section shows, formal risk assessments follow a linear, quantitative path. Risk is defined as a probability of harm times its consequences. Public perceptions of risk, in contrast, are based more on people's attitudes regarding voluntariness, effects on the community, familiarity with the source of the risk, perceptions of fairness in the distribution of risk, and public confidence in the institutions that are managing the risks. The reaction of many experts is that the lay public is not acting rationally, that their intuitive evaluations are not as valid as the methodologically sophisticated risk assessments of the experts. Yet there may be more to the public's perceptions than the experts are willing to recognize. And in a democracy, we need to recognize the validity of these public perceptions when making risk decisions.[13]

HOW DO AGENCIES ASSESS RISKS TO HUMAN HEALTH?

Agencies assess health risks for any of several purposes. Take a decision on setting emission limits for a chemical plant. The prudent course is to determine what substances are in the emissions, how dangerous they are, to how many and what kinds of people, with what effects—in short, to do a risk assessment. For this, we need two kinds of information. The first is on toxicity, a set of measures of the harm of a substance that relates doses to harmful effects—cancer, birth defects, and so on. Toxicity estimates the harm based on different levels and kinds of possible exposures. The second is information on exposure to the substance in the real world. Two chemicals may be equally toxic, but one is emitted in the middle of a city, the other in a lightly populated area. The second poses more risk, in terms of its effects on overall population.[14]

Estimating Toxicity In my example, the first task is to determine whether the emissions from the chemical plant are a hazard and, if so, to estimate their toxicity. At times, this is straightforward, especially in the case of acute hazards, where the link between exposures and effects is direct. An emergency release of ammonia from a chemical plant affects people right away if the dose is high enough, and it is not difficult to link the cause to the effects. For chemicals like dioxin or asbestos, our concern is also with chronic effects—low levels of exposure, over long periods, with a time lag before the effects (e.g., cancer) can be observed. This time lag between exposure and effects is called the latency period. Latency periods are associated most often with cancer, but there may be latency periods for other kinds of health effects as well.

So our decision about the toxicity of chronic hazards is based on scientific evidence about the effects of low exposures over long time periods, often decades. Ideally, scientists would find a group of people who have been exposed to a substance under such circumstances and study them. The field of epidemiology provides tools for such studies. Epidemiologists are concerned "with the patterns of disease in human populations and the factors that influence those patterns."[15] They identify an exposed population, compare it to a control group that was not exposed, account for extraneous factors (smoking or family history), and then draw conclusions about the health effects of exposure to a substance. This kind of evidence is appealing, because it draws on actual patterns of exposure and observed effects on humans.

The problem is that, for a variety of reasons, data on human exposures are unavailable or unreliable. Because exposures are long-term, it is difficult to go back and document patterns and levels of exposure, medical history, lifestyles, and the other information needed for a valid study. Epidemiological data exist mostly for occupational settings, where the exposure levels are higher, records more available, and the range of health effects narrower. For example, data on lung cancer among uranium miners were used to estimate the effects of prolonged exposure to radon.[16] Studies of asbestos drew on the experiences of shipyard workers during World War II. Occupational studies influenced OSHA's decision to set a benzene standard in the early 1980s.[17] Beyond the limited populations that are exposed at high levels in such settings, though, good human data are hard to come by. It often is difficult to extend the very different exposure conditions from occupational to everyday settings, as critics note for radon.[18]

So human evidence is the first choice, but it simply is not feasible for evaluating most chronic exposures. Short of doing clinical studies on humans, the next best method is laboratory tests on animals. Animal bioassays (also called in vivo tests) are the basis for most risk assessments. Scientists administer high doses of substances to test animals, then observe them for tumors or other responses. The responses are extrapolated to draw conclusions about the possible effects on humans. The methods used to extrapolate from test animals to humans vary, depending on what assumptions and models are used. In the 1980s, for example, when assessing the health risks from exposure to

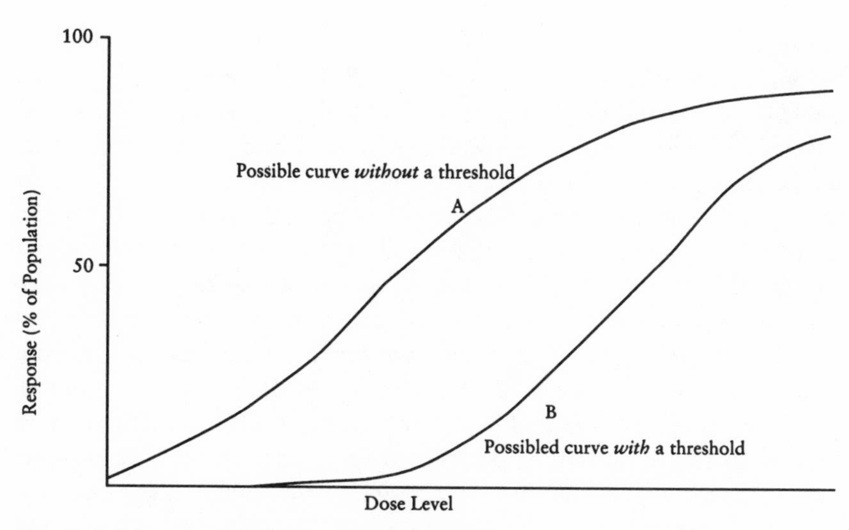

Figure 4. Two simplified dose-response curves.

asbestos, EPA relied on a linear multistage model. This assumes that the likelihood of cancer increases in proportion to the dose and that the tumors form in stages, known as initiation and promotion. Some risk agents (initiators) cause cancer by stimulating changes in cells; others (promoters) support tumor growth once the cell has been transformed to a precancerous state. Asbestos is associated with both kinds of effects, so it is termed a "complete" carcinogen.

The purpose of information on toxicity is to predict the dose-response relationship, to link dose levels with responses in the population. This relationship is presented in simplified form in figure 4. The horizontal axis gives the dose, the vertical axis the predicted response at that dose. A curve plots the relationship between the two, as the percentage of the exposed population that is predicted to show the adverse effect (cancer, birth defects, or others). At times, the curve is smooth; the likelihood of an effect increases in proportion to the dose, as in Line A. At other times, the curve starts out flat, then rises sharply at some level of exposure, suggesting the existence of a "threshold," as in Line B.

Whether there are thresholds below which no effects occur is a major issue. Like other organisms, humans absorb low levels of many substances without ill effects, even substances that are dangerous at higher levels. Indeed, substances like sodium are toxic at very high doses but beneficial at low ones. Many dose-response curves display thresholds, known as no observed effect levels (NOELs), above which we try to avoid exposure and below which we decide there is no harm. For other substances, there are no thresholds, or none that are established, and we assume that any exposure poses some risk.

The policy of government agencies has been to assume that there is no threshold for carcinogens, that any level of exposure may be harmful. In effect, the threshold is zero. Agencies also assume that there is no safe exposure level for suspected causes of genetic damage. For most other effects, however, they have set thresholds, in the form of NOELs or levels of acceptable daily intake (ADI). As the evidence expands regarding the role of different substances in the formation or promotion of cancer, it may be possible to set thresholds for carcinogens, especially those (like dioxin) that scientists view more as promoters than initiators.

Scientists rely on animal in vivo studies not because they are even close to an ideal substitute for human data but because they are the best data available. Scientists use two other methods, short-term in vitro or tissue culture tests and structure-activity analysis, but neither is considered as valid as in vivo tests.[19]

The debate about the validity of animal tests in predicting human effects turns on two critical extrapolations: (1) from high- to low-dose exposures and (2) from test animals (rats or mice) to humans. For chronic hazards, it is not feasible to duplicate the conditions of human exposure in animals. Even if the test animals lived long enough, such tests would be too costly and take far too long. In addition, the rate of cancer occurrence in humans is low enough that it is hard to detect it in animal studies. Lung cancer is the most common form of the disease, but it affects fewer than 8 in 10,000 people annually. Scientists would have to test huge numbers of animals to get a statistical probability of a response (i.e., malignant tumors) for a valid, two-year in vivo study.

To cope with these problems, scientists trade off *time* for *intensity* of exposure. Many studies of chronic effects run for about two years. During this time, test animals are given much higher doses than a human would face in any normal course of events. Scientists usually administer a maximum tolerable dose (MTD), the highest dose test animals can receive without showing effects other than cancer. An issue in the debate over high- to low-dose extrapolation is the evidence that large amounts of many substances may cause changes in cells, making the cells more vulnerable to the formation of tumors. In lower doses, the same substances may have no cancer-causing effects. This has refueled controversy over whether there are thresholds below which some chemicals may not pose cancer risks.[20]

The second extrapolation, from animals to humans, is also problematic. Many differences between test animals and humans are obvious: body weight and size, life span, variations within the species, among others. There also are pharmokinetic differences: metabolism or excretion patterns. Accounting for differences is difficult, given science's limited knowledge about the mechanisms behind the causes of cancer. What makes extrapolation even more difficult, however, is that even among rats and mice, or among animals of different sex within the same species, responses to chemicals vary.

Estimating Exposure Several issues arise in estimating exposure. One is whether to rely on monitoring or modeling. Ideally, personal or ambient monitoring gives direct evidence about exposures. Among personal monitoring methods, biomonitoring takes data from body fluids or tissue samples. More common are ambient data from a given site. They measure the amounts of contaminants people are exposed to but not what is reaching tissues, organs, or cells. Another strategy is to use mathematical models to simulate exposures; these can estimate the movement and fate of pollutants in the atmosphere, surface water, groundwater, and the food chain, for example.[21]

Consider for a moment the task of estimating exposures to drinking water contaminants. In assessing the risks from lead, EPA had to decide what kind of people consumed how much water at what times of the day. For most assessments of exposure to drinking water contaminants, EPA assumed that men consume 2 liters a day, women and children 1.4. For lead, EPA wanted to be more specific. Information on children's intake was especially important, because lead may impair their development. The patterns in consumption also were important; much of the exposure comes from residential plumbing when lead from solder or pipes corrodes and contaminates the water. The longer the water sits in pipes, the higher the lead levels. Levels are highest when water is first drawn from the tap in the morning, or when many hours pass between uses of the tap. To estimate amounts and patterns of consumption, EPA asked a sample of households to keep a diary of when and how much water the adults and children in the household consumed. Based on the diaries, EPA estimated national consumption patterns for the different groups.

In the early 1990s, the special risks that environmental problems could pose to minority communities across the country emerged as a major concern in risk assessment. Some environmental problems affect minorities more than they affect other groups. Lead is one of them; black children who have been tested show elevated blood levels in higher proportions than do white children. Because they are more likely to live in urban areas, blacks and Hispanics also are exposed to more air pollution than are whites. Minorities also are more likely to live near uncontrolled waste sites and to suffer higher risks from pesticides exposures. For a variety of reasons, minority groups often are exposed to higher levels of risks than other groups in the population. In its *Environmental Enquiry* report, issued in 1992, EPA outlined how it would adapt its risk assessment methods to better account for such special risks.[22]

Characterizing Risk Multiplying toxicity times exposure gives an estimate of health risk. The goal in a risk assessment is to describe risk in quantitative terms. The most common measure is excess individual risk, or the increased probability that anyone will experience an adverse effect from an exposure or activity. I noted above, for example, that the risk of someone in the United States dying in a traffic accident in a year was 2 in 10,000 (2×10^{4}). When EPA issued its asbestos regulation in 1989, it estimated the population risk from asbestos in consumer products at about 1 in 1,000,000 (1×10^{6}). For workers exposed at higher doses, the estimate was far higher—from 7 in 1,000 (7×10^{3}) to 7 in 10,000 (7×10^{4}). The first measure described the risks to a person exposed to asbestos at the level typical of the general population; the second measure described the risks to a person exposed to asbestos at the much higher levels typical of certain occupations.

There are several other ways to describe risks. One is to present population or societal risks—the number of cases expected to occur in the population in a year or other time period. EPA estimates that radon in homes causes 7,000 to 30,000 "excess" cases of lung cancer in the United States each year, for example. Another way to describe risk is with statements of relative risk. A 1990 study found, for example, that women who eat red meat daily are two and one-half times more likely to develop colon cancer than are women who eat it only a few times a month.[23] Risks also can be described as a loss in life expectancy; we can say, for example, that smoking reduces the life of a male by 6.2 years compared to nonsmokers.

Two issues are worth noting here. First, agency estimates of health risks tend to be worst-case or upper bound (the upper limit to what is statistically possible). This means that the actual risk is unlikely to be greater and is probably less than the estimates given by agencies. When FDA estimated cancer cases due to saccharin some years ago, it gave the number of expected cases as 1,200. In fact, the analysis gave a range of 0–1,200, but only the upper bound was generally cited in the media accounts. Second, individual and population risks should be seen together. Assume that benzene from a petroleum refinery poses an individual cancer risk of 4 x 10^3 (4 expected cases for every 1,000 people exposed) and that residues of a pesticide on apples pose a cancer risk of 1 x 10^6 (1 for each 1,000,000 exposed). Yet assume that only 1,000 people are exposed to the benzene and most of the U.S. population to residues on apples. The first gives an estimate of 4 excess cancer cases over a lifetime, the second of 200 or more. Conversely, it would be misleading to dismiss the benzene risk because the population looks small; for the 1,000 people exposed, benzene presents a risk.

What do agencies do with these risk estimates? Are there levels of risk at which agencies will always regulate a chemical? Two useful concepts in addressing these questions are *de minimus* and *de manifestis* risk. The first refers to risks too small or trivial to require a response, sometimes described as levels that are "below regulatory concern." The second describes risks so large that they require a response from any reasonable person. An analysis of 132 federal regulatory decisions for which an agency had done a cancer risk assessment found surprising consistency in how agencies responded to cancer risk estimates. Every chemical with an individual risk level above 4 x 10^3 (expected cancers in more than 4 of 1,000 people exposed) was regulated. Except in one case, no chemical posing an individual lifetime risk of less than 1 x 10^6 was regulated. These defined the de manifestis and de minimus risk levels.[24] What of risk levels that fell in between? Here the decision turned on cost-effectiveness, which is examined later in this chapter.

Uncertainty and Risk Assessment Often risk assessment is presented as the technical or value-free side of policy. But the process is full of uncertainty. Assumptions made at many steps may influence results by orders of magnitude.[25] In a study of agency risk assessments, the National Academy of Sciences listed fifty points where agencies had to choose from among scientifically plausible options. At each point, policy views could affect the methods or assumptions and influence the outcome.[26]

Consider the following assumptions that agencies usually make in assessing health risks based on animal studies: that chemicals that cause cancer in animals also do so in humans; that humans and animals are equally susceptible to the effects of substances; that there are no thresholds below which substances do not cause cancer; that human exposures will be the highest that reasonably can be expected; that all substances cause cancer through one mechanism (genotoxicity), which leads them to predict high risk at low doses. Each assumption makes agency risk assessments more conservative, in that they may tend to overstate the actual risks to exposed people.

As a result, two critics of government risk assessment have concluded, agencies rely on "a series of assumptions and policy choices that are designed to overstate the degree of risk posed by carcinogens."[27] These assumptions and choices have a cascading effect, according to critics; each set of assumptions increases the estimates of risk that are obtained in the steps that follow.

EPA's policy is illustrative. In its *Unfinished Business* report, the agency described the assumptions it made in evaluating cancer risks. One was that human sensitivity to a substance is as high as the most sensitive animal species. Another was that benign tumors should be counted as heavily as malignant ones in estimating cancer risk. A third was that the agency would rely on an upper bound estimate (the upper limit to what is statistically possible) of risk rather than a more realistic middle range estimate. The upper bound gives a very high estimate of likely risk; the true risk is unlikely to be higher and will probably be significantly lower. Although not "a realistic prediction of the risk," EPA observes, "it is a reasonable precaution in the absence of more knowledge of the mechanisms behind cancer."[28]

So at nearly every step in a risk assessment, the policy of most agencies is to make the most protective assumption or policy choice. Agencies cope with uncertainty by being conservative in their choice of methods; if there is error, it is in the direction of overestimating, not underestimating, risk. The principle is that it is prudent public policy to assume the worst when there is uncertainty about the scientific basis of decisions about health.

Frances Lynn's study of relationships between the organizational affiliations of scientists and the assumptions they make in risk assessments sheds light on the role of values in risk assessment. In a survey of occupational health scientists in government, universities, and industry, she concluded that "there were links between political values, place of employment, and scientific beliefs." She found that "scientists employed by industry tended to be politically and socially more conservative than government and university scientists."[29] Industry scientists were more inclined to make choices or adopt assumptions that led to lower risk estimates. Government experts were most protective in their choices and assumptions; university experts were between the two. Industry experts were more skeptical of using animal tests to predict human risk and of the "no threshold" assumption for cancer; the other groups of scientists were more inclined to accept both.

My point is not to discredit accepted techniques for risk assessment. For all their limitations, they are the best tools available for evaluating the potential hazards of a variety of agents when human data are lacking. But in using the results of these studies, we should keep three points in mind. First, we cannot neatly isolate risk assessment as the purely technical, value-free side of policy. Even scientific analysis requires policy choices. Second, government risk assessments deal with uncertainty by adopting very cautious assumptions that tend to overestimate rather than underestimate potential risks. This may be a prudent policy, but it does introduce certain biases into decisions. In what they call the "perils of prudence," two critics argue that government risk assessment policies bias agencies toward regulating for cancer and away from regulating for other, more serious health risks.[30] Third, because of the uncertainty involved in risk assessment, it may make sense to use it more as a rough guide for setting priorities and comparing problems and less as an exact rule for decisions. I develop this point more in chapter 5.

At the same time, there are many reasons to think that risk assessments may not be conservative enough. They often focus only on one route of exposure for a given pollutant, not all the likely routes of exposure. Single-medium risk assessments may not account for the cumulative effects of multiple sources of exposure to, say, asbestos or lead. Different pollutants can have synergistic effects—interactions among them that make the overall risks more serious than just the sum of the risks of individual pollutants. In addition, groups in the population may be especially sensitive or vulnerable to different kinds of risks, requiring even more conservative assumptions in risk assessments. For example, in 1993, a panel of the National Academy of Sciences recommended that agencies modify their risk assessment practices to account for the special effects on children of pesticide residues on foods.[31]

ECONOMIC ANALYSIS AND THE ENVIRONMENT

Risk is part of the analytical basis for policy. Economics, specifically, the analysis of costs and benefits, is another part.

Why analyze costs and benefits? Because nothing is free in life, in environmental policy as in anything else. One of the premises defined in chapter 1 was that when we take action to deal with a problem, we give up benefits that we could have achieved by allocating the same resources to another purpose. Resources that are allocated to solving environmental problems *could* have gone to another social goal—education, defense, or health care.

When government agencies spend money from their own budgets, the options are presented and the choices made in the annual budget process. Money that is not spent for education may go to defense or health care, according to the appropriations worked out by the government. When agencies issue regulations that direct private firms or other levels of government to spend money for some purpose—to treat drinking water or take pollutants out of water discharges—they are allocating society's resources, just as they do in the budget process. The difference is that by allocating society's resources indirectly through regulatory requirements rather than directly through their own budgets, agencies do not have to account for resources or trade them off in the same way.

So the problem, to many people, is that agencies are not held accountable for the costs they impose on society when they issue regulations. For this reason, many people have called for the adoption of a regulatory budget that limits the costs that regulations can impose on society and forces agencies to trade off competing goals and the resources needed for achieving them.[32] The United States has not adopted any such regulatory budget proposal, in large part because of the practical difficulties of estimating costs with precision. Instead, critics of the costs of government regulation have tried to hold agencies accountable by making them analyze (and thus justify) the costs and benefits of their regulations.

THE CONCEPTS OF COSTS, BENEFITS, AND DISCOUNTING

The concepts of costs and benefits are intuitively simple but analytically complex. At a commonsense level, most of us think easily in terms of costs and what we are getting in return for them (benefits). But translating these commonsense concepts into what analysts call "operational" categories for making decisions is another matter, even more so with benefits than costs.

Let us begin with simple definitions. E. S. Quade's book on policy analysis provides a starting point. Costs, he observes, "are the negative impacts associated with a decision."[33] We can distinguish several kinds of costs—direct or indirect, capital or operating, average or marginal, to follow some common distinctions. Direct costs are easiest to account for, because they are related most clearly to a decision. Making a utility spend $125 million to install a scrubber that removes sulfur oxides from its emissions poses direct, identifiable costs. Some of these are the costs of designing, purchasing, and installing equipment, known as capital costs; others, servicing equipment or buying energy to run it, are known as operating costs. There also are indirect costs, like higher prices for consumers.

A key concept is that of marginal or incremental costs. These are the costs that are incurred when we move from one level of control to another (usually more stringent) one. Unless we are addressing sources for which there are no requirements in place, our focus is usually on marginal costs, which tend to increase as controls become stricter. Indeed, a nearly universal principle for environmental managers is that the costs of taking each additional unit of pollution out of a waste stream increase, sometimes dramatically, as they try to approach a zero discharge level.[34]

So the most obvious costs are the direct costs of compliance to companies or other targets of regulation.[35] Most of these costs fall on the private sector, but some fall on government. However, most programs impose costs beyond the direct costs of compliance. To account for these, we need to think in terms of social costs, or the value society places on the goods and services that are lost when resources are diverted to environmental ends. In addition to direct costs, they include implementation costs, the adjustment costs for displaced resources (such as lost production capacity in industry and transition costs for displaced workers), and losses in producer or consumer surplus. Regulation may cause a decrease in the output of goods and services in an industry. When EPA cancels the registration for a pesticide, it makes farmers use substitutes or (if none are available) not use a pesticide at all. If a substitute is less effective, or there is no substitute, farmers experience a loss in yield that they cannot make up, at least in the near term. This loss in producer surplus is a social cost.

Benefits are the other side of the coin. We can define benefits as the desirable or favorable effects associated with a decision. One way to see them is as costs avoided. By reducing sulfur oxide emissions from utilities, for example, we avoid the costs that society would have had to bear from the higher emission levels. These include health problems, acidification of lakes or forests, and impaired visibility in parks and scenic locations. We could further break down the health costs that are avoided by accounting for illnesses (lost workdays or measures of impaired health) and mortality (premature deaths associated with exposure). Benefits extend well beyond the avoidance of unwanted health effects, however. We all enjoy the nonhealth benefits of environmental programs. Some have recreational or aesthetic value: fishing in unpolluted streams, swimming in clean lakes, viewing the Grand Canyon, or enjoying the natural variety of life in a wetland. Other benefits are more tangible and commercial: fisheries, healthy crops, or productive forests with commercial value.

It is one thing to list the categories of benefits we expect from a program. It is quite another to assign numerical values to those benefits. Why worry about assigning numbers to expected benefits? The short answer is, to enable us to compare them to costs through a common metric. That metric is dollars.

How do analysts assign dollar values to benefits? Methods and principles for valuing benefits come from welfare economics. The key concept is that of people's "willingness to pay," measured by observing behavior or conducting surveys.[36] Economists assume that by summing up individual tastes and preferences in society, we can derive an appropriate measure of the economic value of a resource. The most direct way of determining preferences is through markets. To the extent that environmental problems affect the amount or quality of goods or services that are traded in markets, economists can assign value to them. "Markets reveal value," as OMB once put it.[37] If we know that pollution reduces the amount of shellfish that is harvested in an estuary, we use the market value of shellfish to determine the benefits of reducing pollution.

But what of resources that are not traded in markets? Wetlands offer many benefits to society: they provide habitat for many forms of wildlife, filter pollution out of water, and have aesthetic value. But wetlands are not traded in markets. Nor are many other values that are protected by environmental programs—endangered species, a view of a mountain peak, even human life.

For these "nonmarket" goods, analysts determine preferences with indirect methods—by observing behavior or with surveys that assess people's willingness to pay to preserve resources or improve environmental quality. Several kinds of behavior give information about preferences. One approach to valuing life, for example, is to compare the wages people are paid to perform more and less risky jobs and then measure the differences. These comparisons measure how much people want in compensation for performing higher-risk jobs. Analysts often use a method known as contingent valuation (CV) to assign values to goods that are not traded on markets. People are asked in a survey what they would pay for an incremental change in environmental quality. Based on their responses to a series of questions about their willingness to pay for specific kinds of gains in environmental quality, or to accept losses in environmental quality, analysts assign a dollar value to those gains or losses. The validity of the surveys and CV generally is a matter of debate, although many economists argue that they are more reliable than other indirect methods of valuing benefits.[38]

Consider the task of valuing health benefits. They come in two kinds, illnesses avoided (morbidity) and deaths avoided (mortality). Avoiding illness reduces demands on the health care system and avoids days of lost work or diminished productivity. These have an economic value that gives a measure of their monetary benefits to society. If we can describe the number and kinds of illnesses that are avoided, we can estimate the health care costs and productivity losses we avoid by reducing pollution.

But what is the value of an avoided death? In the guidelines for cost-benefit analysis that it issued in the early 1980s, EPA cited several studies that measured the levels of compensation people appeared to want to perform riskier jobs. These studies found that annual wages were $4 to $70 higher for jobs that exposed people to a 1 in 100,000 greater risk of death. This translates into a value of $0.4 to $7 million (in 1982 dollars) for each statistical life saved. A more recent review of studies on the willingness to pay for reductions in the risks of death found that people would pay from $1.6 to $8.5 million for each death avoided.[39]

Earlier, we discussed an analysis of 132 cancer risk decisions made by EPA and other federal regulatory agencies. It concluded that agencies always regulated when individual risks were above 4×10^3 and almost never regulated when risks were below 1×10^6. Between these de manifestis and de minimus levels, the determining factor appeared to be the cost of regulation. Agencies regulated when the cost was less than $2 million per life saved but not when the cost was higher. The $2 million figure was not explicit in the policies of any of these agencies, but it was the implicit value reflected in cancer risk decisions.[40]

Costs and benefits are two key concepts. Discounting is a third. Costs and benefits do not accrue immediately. Analysts think in terms of the stream of costs and benefits that flow over time. This means that we cannot value the effects of a decision without accounting for time. Spending a million dollars today is in real terms a much greater cost than spending the same million in twenty years. If we spend in the future, we have use of the money now for other purposes. By the time we spend the money, inflation will have reduced its value. To compare money spent down the road with money spent now, we discount future costs to their present value. To calculate present value, analysts apply a discount rate to the stream of future costs; the higher the discount rate, the more the time that has passed, the less is the present value of the costs that are expended or the benefits that are realized. There is more controversy about discounting costs than benefits. Will a wetland and its functions be worth less in thirty years than they are today? Even at a low discount rate (say 3%), the value of the wetland will turn out to be very small if we discount its benefits over a period of thirty years.

THE COSTS AND BENEFITS OF ENVIRONMENTAL PROGRAMS

What do we know about costs and benefits? In 1990, EPA completed a comprehensive analysis of the costs of environmental programs in the United States.[41] In this analysis, the costs are annualized and adjusted to 1986 dollars to make them comparable over time. I use three years for comparisons: 1972, when the national air and water programs began; 1987, the most recent year in the report for actual (not projected) data; and the projections for 2000.

In 1972, the total annualized costs for all pollution control in the United States were $26 billion, which was 0.88 percent of the GNP for that year. By 1987, the amount had risen to $85 billion and 1.92 percent of the GNP. For 2000, costs are projected to rise to $160 billion (2.83 percent) of the GNP. So in less than thirty years, the total costs in billions of dollars will have risen from $26 to $160 in real (1986) dollars and the percentage of the GNP from under 1 percent to nearly 3 percent.

Table 7 gives a summary of the trends by environmental medium. Here are some of the highlights for each medium (in billions):

> The United States spent $26.7 for air quality in 1987. This will rise to $44 by 2000 under the 1990 CAAA. (The higher numbers in the table include radiation program costs.) Estimates for the CAAA's provisions for ozone, acid rain, and air toxics will amount to roughly $14.6 of the total by 2000.
>
> The United States spent $34.4 for surface water quality in 1987. This will grow to $57.6 in 2000. Drinking water treatment cost $3.1 in 1987 and will amount to $6.6 in 2000. Costs associated with point sources accounted for 90 percent of the 1987 costs for surface water quality.
>
> The United States spent $19.1 for solid and hazardous waste programs in 1987. This is up from $8.4 in 1972 and will grow to $46.1 in 2000. These costs include solid waste collection/disposal. Of the 2000 estimate, some $38 is associated with the RCRA and $8 with the Superfund program.

On a different scale from the air, water, and waste media are the costs of pesticides and toxics regulation. These averaged $0.68 billion annually from 1981 to 1988, but they are expected to rise to $2.9 billion in 2000, when private sector pesticide costs will reach $1.7 billion. In addition, there is a small category of "multimedia" costs that cannot be allocated to a single medium (energy and basic research, management and support, and community right-to-know programs).

Of the three main categories of media costs (air, water, and land), those for land will rise the most in real ($27 billion) and percentage terms (240%) between 1987 and 2000. Air and radiation costs will increase by just under 170 percent and water costs by just over 170 percent over this same period. The increases in the land category are largely a result of the implementation of hazardous waste regulations.

Whether dedicating 2 to 3 percent of the GNP to the environment is

TABLE 7 COSTS OF ENVIRONMENTAL PROGRAMS IN THE UNITED STATES
In Billions of 1986 Dollars

Medium/Program	*1972*	*1987*	*2000*
Air and radiation	7.9	27.0	44.9
Water	9.9	37.5	64.1
Land (waste programs)	8.4	19.1	46.1
Chemicals (pesticides and toxics programs)	0.1	0.8	2.9
Multimedia	0.1	0.8	2.3
Total	26.5 (0.88% of GNP)	85.3 (1.92% of GNP)	160.4 (2.83% of GNP)

SOURCE: *Environmental Investments: A Summary* (Washington, D.C.: U.S. Environmental Protection Agency, 1990), 2-2 and 2-3. The numbers are rounded to the nearest tenth billion. Estimates for the year 2000 are based on the "full implementation" scenario.

too much or not enough is a matter of judgment. Any society has to decide how much to allocate to competing objectives. It is useful to compare environmental spending with spending for other kinds of goods and services. In 1987, to take some examples, while the United States was spending just under 2 percent of the GNP on pollution control, it was spending 4.2 percent on clothing and shoes, 6.9 percent on national defense, 7.0 percent on medical care, 9.3 percent on housing, and 11.7 percent on food.[42]

Another way to think about U.S. spending on the environment is to compare it to what other countries spend. I will use 1985 for comparison, as data for this year are available for several European countries. In 1985, the United States spent 1.44 percent of its gross domestic product (GDP) on pollution abatement and control (GDP differs from GNP in netting out certain kinds of foreign trade effects). This was more than what was spent in all but one of the other countries, including West Germany (1.52 percent), Finland (1.32 percent), the Netherlands (1.26 percent), the United Kingdom (1.25 percent), France (0.85 percent), and Norway (0.82 percent). These offer rough comparisons. To make them truly comparable, it would be necessary to consider the pollution levels in each country and the relative efficiency of each country's pollution programs. Put simply, one country may get more environmental protection with a smaller percentage of GDP if its programs are more efficient. But these figures at least give us a basis for comparison.[43]

Another method for deciding how much of society's resources should go to the environment is to compare costs and benefits. One study, for example, estimated the total benefits of programs to reduce point source water pollution in 1985 at $5.7 to $27.7 billion, with $14 billion as the most likely estimate.[44] This study included four kinds of benefits: recreational use, accounting for half the total; nonuser benefits (those from aesthetic or ecological gains in water quality); improved productivity in commercial fisheries; and the benefits of diversionary uses of water (i.e., water used for drinking or industrial purposes).

In comparison, EPA estimated the costs of all surface water quality programs in 1987 at $34 billion. This is unfavorable when compared to the $14 billion in estimated benefits. But remember that the cost estimates cover *all* surface water quality programs, while the benefits estimates do not include newer controls on metals and toxics or the benefits of control programs for nonpoint sources of water pollution. The benefits estimates also are two years older. Even if we doubled the benefits estimates to account for such factors, however, there still would be a gap between the benefits (doubled at $28 billion) and the costs (at $34 billion).[45]

More recently, Paul Portney estimated the costs and benefits of the Clean Air Act Amendments of 1990.[46] These are incremental costs and benefits that will occur on top of the $27 billion that was spent on air quality in 1987. Portney analyzed the three main parts of the CAAA: the controls on utilities' sulfur oxide emissions to reduce acid deposition; the limits on industrial emissions of volatile organic chemicals and the stricter tailpipe standards on mobile sources, mainly to reduce urban ozone; and the requirements that major sources of about 190 toxic chemicals install Maximum Available Control Technology to reduce emissions.

The return on the sulfur oxide controls looks good in the analysis, with costs at about $4 billion and benefits at $2 to $9 billion. Drawing on several studies, Portney estimated the costs of the urban ozone provisions at $19 to $22 billion annually by 2005. Benefits in the form of avoided health costs, higher agricultural productivity, reduced damage to forests and materials, and improved visibility came to $4 to $12 billion per year. The costs of the toxics controls were hardest to estimate (the MACT standards were yet to be set) but were set at $6 to $10 billion. The benefits of the toxics limits ranged from zero (given the chance that nobody may be exposed to harmful levels) to about $4 billion per year. Based on this analysis, the annual incremental costs of the CAAA would amount to some $29 to $36 billion, the benefits to $6 to $25 billion.

What of the costs and benefits of U.S. environmental programs as a whole? Drawing on several studies, Robert Hahn and John Hird estimated that the annual costs of environmental programs in this country ranged from $55.4 to $77.6 billion and the annual benefits from $16.5 to $135.8 billion, with $58.4 billion as a "most likely" estimate (in 1988 dollars). On the whole, they conclude, the benefits appear to slightly exceed the costs. Still, they urge that the incremental costs and benefits of new programs be evaluated carefully, because each unit of pollution reduction from here on may be more expensive relative to the benefits than has been the case to this point.[47]

Many uncertainties surround these estimates of costs and benefits. The range in the Hahn and Hird estimate of benefits ($16.5–$135.8 billion) conveys the uncertainty in the benefits side of the picture. And there is dispute about the assumptions and methods used to generate these numbers, especially regarding the benefits. But they do provide a rough guide to what American society gets for its investments in environmental programs. They suggest that although such programs offer substantial benefits, they impose large costs as well and that the incremental costs of some new programs may exceed the likely benefits. If nothing else, these studies stress the value of using economic analysis to inform policy makers when they design new programs.

ECONOMIC ANALYSIS IN ENVIRONMENTAL POLICY DECISIONS

Since 1970, economic analysis has played an important yet changing role in policy making. This section reviews three forms of economic analysis that EPA and other agencies have used in making policy decisions: economic impact analysis, cost-benefit analysis, and cost-effectiveness analysis.

Economic Impact Analysis At a conceptual level, impact analysis is common sense, something we do intuitively all the time: If we do something, what will its consequences be? An early use of the impact model of analysis was for the Environmental Impact Statement that was required under the NEPA. The impact model was applied to other effects in the 1970s as well, for example, in the Urban and Community Impact Analysis and the Regulatory Flexibility Analysis. The intent with each was to make agencies evaluate and thus account for a subset of the consequences of their actions.

Aside from the EIS, the most important use of impact analysis has been to predict the economic effects of government actions. Environmental regulation has been a special focus.[48] The first general requirement came in 1971, when President Nixon directed agencies to prepare "Quality of Life" reviews that summarized the objectives, alternatives, costs and possible benefits, and other effects of proposed decisions. These were simple analytically; their purpose was more to allow OMB to review rules than to advance the state of the art of analysis. Later, President Ford directed agencies to prepare Inflation Impact Statements. This reflected the pressing concern of the day, the rising inflation rate.

The purpose of economic impact analysis is to predict the effects of agency actions on companies, industries, economic sectors, or the economy as a whole. It includes several kinds of effects. Some are effects on firms: How will a rule affect a company's ability to raise capital? Will it increase the price of raw materials? Will it force firms to close? Other effects are those on consumers or economic sectors: Will workers in affected industries lose their jobs? Will American companies have trouble competing with companies overseas?

Impact analyses posit a relatively simple cause-and-effect model. If we take action X, then we can predict consequence Y. This is not to say that the analysis is simple or the need for information insubstantial. A defensible economic impact analysis is a demanding piece of work. As one moves from the direct or immediate effects of decisions (the costs of production or job losses in affected firms) to indirect effects (consumer prices or job losses in the community), the analytical difficulties grow.

Cost-Benefit Analysis Impact analyses examine only the cost side of the equation. What are we getting in return? Do the benefits we can expect from a decision justify those costs?

To answer such questions, we need to do two things. One is to broaden our definition to include *social* costs. The other is to describe and value the *benefits* that accrue to society as the result of a policy. This takes us to cost-benefit analysis, whose conceptual foundations were discussed earlier in the chapter.[49]

In very simple terms, we can say that cost-benefit analysis proceeds in four steps.[50] The first is to identify the expected effects on society and categorize them as costs or benefits. The second step is to assign values, usually in the form of dollars, to each of these categories. The third step is to discount costs and benefits to account for the effects of time. Because we expect both the desirable and undesirable effects of a decision to occur in a stream over time, it is necessary to discount them to their present value to make them comparable. The fourth step is to calculate the ratio of the costs to the benefits and select "the combination of policies that maximizes net benefits."[51] All of this, it should be obvious, is much more easily said than done.

A few years ago, EPA issued a report on its use of cost-benefit analysis between 1981 and 1986.[52] Over that period, it prepared fifteen Regulatory Impact Analyses, at an average cost of almost $700,000, which EPA noted was about 0.1 percent of the minimum cost of a major rule over five years. Although cost-benefit analysis is usually seen as a way of limiting government regulation, the report noted that in one case the RIA led EPA to adopt a stricter standard by eliminating the use of lead in motor fuels. The RIA revealed benefits in the form of reduced health effects, medical care, and educational costs for children with high blood lead levels; fewer deaths, illnesses, and lost wages from cardiovascular and other diseases; lower emissions of pollutants besides lead (as a by-product of the lead controls); and gains in fuel economy and lower vehicle maintenance costs. The present value of benefits over costs (i.e., the net benefits) came to $6.7 billion for the period 1985–1992, a finding that clearly supported the decision to lower and later ban lead in fuels.

Cost-Effectiveness Analysis The easiest way to grasp the notion of cost-effectiveness (c-e) analysis is to view it as a truncated form of cost-benefit analysis. Knowing how social costs compare to social benefits tells us something about what our policy goals should be. If the benefits of reducing lead levels in drinking water are shown to exceed the costs by a ratio of three to one, for example, there is an analytical basis for selecting lead reduction as a policy goal. Often, however, the goal is already decided on, or it is infeasible to derive or use estimates of benefits. In such cases, c-e analysis is another analytical tool. It is less comprehensive than a full cost-benefit analysis but broader conceptually than economic impact analysis.

Take controls on air emissions as an example. For areas of the country that have not attained the NAAQS, the goal is clear: Get emissions down to the level necessary to meet the standard, whatever the costs. Congress set this goal when it told EPA to set the NAAQS at a level that protects public health. The policy goal is set. Where there is room for choice and for the use of analysis is in selecting a cost-effective mix of controls. Volatile organic chemicals are a principle ozone precursor and are emitted by motor vehicles, refineries, manufacturing plants, even lighter fluids. To devise a strategy for VOCs, we can estimate the cost per ton of reducing emissions from different sources and select those that are most cost-effective, those that cost the least per unit of reduction.[53]

The cost per unit of pollution reduced is one measure of cost-effectiveness. There are others. Policy makers compare options based on the number of statistical lives saved (or deaths avoided) as a result of a policy. Illnesses or other avoided effects are a common basis for comparison. Nearly any class of environmental effect can provide a basis for comparing cost-effectiveness.[54]

Evaluating the Different Forms of Analysis Few analytical issues have been as controversial as cost-benefit analysis. Some criticisms relate to the methods, such as the problems in assigning a dollar value to benefits. Others are ethical, such as accounting for the distributional effects of decisions (on minority groups, for example) in an approach stressing aggregate social benefits. Others criticize the technocratic biases of using expert analysis and quantification to make major policy decisions. Robert Zinke argues that if they rely too much on cost-benefit analysis, policy makers may reduce the opportunities for public debate and allow experts to manipulate their decisions. He is critical of the "arbitrariness with which cost-benefit criteria are imposed,"[55] a point that could especially be applied to OMB in the Reagan/Bush years, when analysis often was used to justify predetermined policy goals rather than to make more rational or genuinely cost-beneficial decisions.

Cost-effectiveness analysis overcomes some of the criticism. It allows policy goals to be set through other mechanisms, which may include a comparison of costs and benefits, while informing policy makers of the most efficient way of achieving them. It offers a flexible tool for comparing the costs of policy options in whatever terms policy makers wish to use—by units of pollution reduced, lives saved, illnesses avoided, and others. There is disagreement about using cost-effectiveness analysis but more agreement that it offers a valuable tool for informing choices, with fewer of the problems associated with cost-benefit analysis.

Traditional cost-benefit analysis is so full of weaknesses in assumptions, so weighted down with uncertainty, and so discredited by the ideological biases of the Reagan/Bush administrations that it may not be the appropriate analytical model for the coming years. But this does not mean that risk analysis and economic analysis should not be a valuable and necessary part of environmental policy making.

My own view is that cost-benefit analysis is most useful in setting broad priorities, in defining which problems deserve more or less attention. Knowing that the likely benefits of reducing lead exposures exceed the costs by a ratio of three or four to one provides valuable information for policy makers. Similarly, if an analysis demonstrates that the potential benefits of controlling nonpoint source pollution in an area greatly exceed the benefits of imposing further controls on point sources of water pollution, it suggests that we should give our attention to the first over the second. But in selecting an approach to reducing lead exposures or to controlling sources of nonpoint water pollution, it may be more valuable and defensible to evaluate the relative cost-effectiveness of the policy options and select the ones that give the best return (in terms of costs per life saved or other damages avoided) for the money that is spent.

ISSUES IN THE USE OF RISK ANALYSIS AND ECONOMIC ANALYSIS

The use of analysis in environmental policy making raises several issues. Here are three: Can environmental agencies affect the cancer rate? Does cost-benefit analysis undervalue ecological resources? Should a dollar value be assigned to human life?

CAN ENVIRONMENTAL AGENCIES AFFECT THE CANCER RATE?

The irony in the almost exclusive emphasis on cancer for many years is that agencies like EPA may be able to have only a marginal effect on the cancer rate. In a recent analysis of EPA and FDA risk assessments, Michael Gough concluded that EPA regulations would, at best, prevent some 6,400 of the 485,000 annual cancer deaths in the United States (1.3 percent of the total).[56] The number of preventable cancers is even smaller if we rely on FDA's less conservative methods for risk assessment, where an estimate of preventable cancers would be about 1,400 (0.25 percent of the total). Both of these estimates assume that EPA regulates all chemicals that may be carcinogenic, so they give the highest reductions in risk that EPA could hope to achieve.

How did Gough reach these conclusions? Bear in mind first that only 2 to 3 percent of all cancers are associated with environmental pollution, "in the sense of exposure to contaminated air, water, and soil."[57] Another 3 to 6 percent are associated with radiation, only a small part of which is affected by EPA's regulations. So the effect of environmental regulations is limited to a small fraction of total cancer deaths. Gough compiled EPA and FDA estimates of cancer risks for several environmental problems, from stratospheric ozone depletion and indoor radon to drinking water contamination and hazardous waste sites. He estimated the total cancer deaths associated with the list of environmental exposures at 22,000 to 44,000 annually (17,000–39,000 with the FDA method), out of the total of 485,000 annual cancer deaths in the United States.

Still, of this 22,000 to 44,000 cancer deaths, only a maximum of about 6,400 could be affected by EPA regulations (1,400–1,600 with the FDA method). Some risk sources fall outside EPA's legal authority, such as workplace chemicals and consumer products. Others are only partly affected by EPA's actions, for example, stratospheric ozone depletion, natural radiation, and tobacco smoke. Among the exposures that EPA's regulations may affect are those from pesticides on food, hazardous air pollutants, inactive hazardous sites, and drinking water, which together account for nearly 80 percent of the cancer risks EPA has the authority to deal with. However, EPA and other agencies reduce risks like tobacco smoke, radon, or ozone depletion through means other than regulation—with public education or state and local restrictions on smoking in public places. If agencies are to have any chance of significantly reducing cancer rates, it will be by influencing diet and behavior, not by regulating pollution. One review on cancer concludes, "The widespread public perception that environmental pollution is a major cancer hazard is incorrect."[58]

The key message is that environmental programs can reduce the cancer rate only marginally. Whatever the causes of cancer, they appear to be largely out of EPA's control, at least under its current authority. This does not mean that cancer risks are an inappropriate target for government action. When there are high exposures of special populations (cancer hot spots), government action may reduce risks that look small in the overall national figures but are significant locally.[59] It does, suggest, however, that a focus on cancer over other kinds of risks is misplaced. It supports a recent trend by policy makers to give more attention to noncancer health effects—reproductive, neurological, and others.

DOES COST-BENEFIT ANALYSIS UNDERVALUE ECOLOGICAL RESOURCES?

I remember a children's book from the late 1950s about a New York City garbage collector who enjoys knowing that the trash he collects will be barged to New Jersey to fill in swamps. Views about swamps have changed in the last thirty years. We now call them wetlands and value them as resources worth preserving. This is partly because the acreage of wetlands has diminished (there is greater value in scarcity) and partly because we know more about the crucial ecological functions that wetlands perform. Our views about wetlands have changed; we value them more than the New York City garbage collector and his contemporaries did decades go.

How we value ecological resources over time depends in part on how we design the analyses we use to make decisions. In *Reducing Risk,* EPA's Science Advisory Board criticized the way cost-benefit analysis was being used to evaluate ecological risks, especially in terms of discounting.[60] Discounting is a way to account for time and risk in comparing costs and benefits in the future. It is hard to argue with the fact that money spent in the future has less value than money spent today. But what about ecological benefits? Are they worth less if they are realized in thirty years rather than now?

SAB had three criticisms of discounting. The first is based on the notion of "intergenerational equity." Resource valuations are based on people's willingness to pay. When people vote their preferences in markets or surveys, they make choices for future generations who lack a vote. But why do we assume that this generation should make choices for future ones? Second, SAB argued, discounting undervalues the harmful effects of "large-scale and long-term" problems. The benefits of policies that mitigate global warming will be felt well into the future; their costs will be borne much sooner. It is difficult for benefits that will accrue well into the future to compete with costs that are incurred now, especially when those benefits are discounted to their present value. Third, discounting treats ecological resources the same as physical and human capital—as exchangeable commodities. But ecological capital cannot be substituted for other kinds of capital; species diversity or wetlands cannot be bought and sold like machinery or computers, and they should not be treated in the same way.

Among some economists, there is a movement to treat ecological capital as "special goods."[61] They argue that investing in the environment is more akin to investing in health or education than in computers or a new factory. As Frances Cairncross points out, a discount rate that is right for consumers or companies "may not be right for society as a whole. If it were, not a single school might be built, since the return on investing in a child's education would be too low compared with that on junk bonds."[62] For long-term or large-scale risks, "it may be more appropriate to adopt a zero discount rate," SAB advised.[63] At least as it is currently practiced, it is fair to say that cost-benefit analysis may significantly undervalue ecological resources.

SHOULD A DOLLAR VALUE BE ASSIGNED TO HUMAN LIFE?

The study of 132 regulatory decisions discussed earlier concluded that $2 million appeared to be a cutoff that agencies used in deciding whether to regulate middle-range risks. The studies of willingness to pay assigned a value of $1.6 to $8.0 million for each death avoided. Based on such studies, policy makers might value a life at $2 to $3 million and use this as a decision rule for making regulatory choices. That is, they should regulate when the cost per death avoided is under that amount; they should not regulate when the cost is above that amount.

It is at this point that people begin to feel uncomfortable. Here we face a conflict between the rational aims of risk and economic analysis and the symbolic and moral side of policy. It is difficult to accept an "exchange rate" for ecological resources; it is even more difficult to accept an exchange rate for human life. Should a sacred value like human life be treated as an economic commodity?

One response is to say that environmental decisions inevitably involve trade-offs in which we attach value to lives. Risk and cost analyses merely make these trade-offs explicit. We deal with statistical lives in these analyses, in the form of predictions about the likely health risks to exposed populations. There is no one person whose health or life is threatened, though there may be exposed populations in which risks are high. If we value life, should we not maximize the deaths that are avoided for any expenditure of society's resources? When a state legislature determines a budget for prenatal health care, it affects infant mortality rates, and it implicitly assigns value to life. Similarly, when health agencies reduce the insurance coverage for serious illnesses, as a way of rationing scarce health care dollars, they affect the availability of care and people's chances of survival. Should not these choices be explicit, in environmental, health, or any other area of policy?

Another, contrasting point of view is that even assigning implicit value to human life is morally questionable. Ethically and in the public mind (as revealed in risk perception research), there is more to the concept of risk than just the number of statistical deaths or other harmful effects. We also should account for the sources of risk, whether the exposures are voluntary, the kinds of people that are exposed, and where the risk is concentrated. Factors besides mortality enter into decisions: other health effects; damages to soil, crops, and forests. In making these choices, policy makers should not substitute absolute decision rules for judgments that can take all of these factors into account.

There is something to be said for this second argument. It may be, as Douglas MacLean observes, that in acting rationally agencies should not be too open, too absolute, or too public. "Like it or not," he comments, "the actions of EPA have important symbolic and expressive significance."[64] Even when costs play a central role in decisions—and there is not agreement that they should—it may be best not to be explicit about where the lines are drawn, in terms of the value of a human life. As long as costs and risks are analyzed, policy makers will attach implicit value to lives. Yet so far, agencies have not adopted explicit guidelines on what those values should be. There are good reasons for their reluctance.

ANALYSIS AND THE ENVIRONMENT

Risk analysis and economic analysis are now almost routine at many environmental agencies, yet the debate over when and how much to use analysis is a lively one. Several factors promote greater use. One is simply having the information needed to make sensible choices. If there are an almost unlimited number of sources of exposure to potentially toxic chemicals, we need to know which among them threatens health or the environment the most. We can do this by assessing relative risks. Similarly, in deciding on a control standard for a pollutant, we may want information about the costs and effectiveness of different controls to determine what level of stringency to require to get the most for the costs.

Legal factors also promote a greater use of analysis. When there is a legal requirement to justify regulation on cost-effectiveness or cost-benefit grounds, as in the TSCA, agencies carry a legal burden that can only be met by analysis. There are political reasons for analysis as well. If an agency proposes a rule that will impose large costs, require changes in behavior, or otherwise disrupt people's lives or livelihoods, it has to be able to justify that rule. Executive orders requiring various forms of economic analysis reinforce these political factors.

At the same time, there are formidable obstacles to the greater use of analysis in making environmental policy. The most compelling one over the last decade has been the trend toward prescriptive environmental laws that leave very little room for agency discretion and thus for using analysis to inform choices. The more Congress directs EPA to issue standards regulating a set list of pollutants, or to base those standards on strict definitions of the "best available technology," the less likely it is that risk and economic analysis will play a role in the outcome. Under the Reagan and Bush administrations, analysis acquired a bad name. Perhaps the Clinton administration will be able to do better.

Another obstacle is limited resources. A reasonably thorough cost-benefit analysis for a major rule costs a million dollars or more. At a time when the list of issues is expanding and EPA's analytical resources are shrinking, analysis must be used selectively. Adding to this are limits on policy makers' time, especially given the tight deadlines that Congress has set for many regulations. The number and range of choices that have to be made also set constraints, as agencies struggle to link available analyses with the decisions *when* they have to be made.

A rational model assumes complete and reliable information, an almost leisurely pace for making decisions, agreement on goals, and broad discretion that allows rational choices to be made. As the last three chapters show, such conditions rarely are met in the world of environmental policy. Different statutes set multiple and sometimes conflicting goals. Congress, the White House, and the courts pull policy makers in many directions, with industry and environmental groups contributing to much of the pulling. Within agencies, choices are fluid. Options come and go, data are more or less reliable, the influence of policy advocates in the agency rises and falls. Agencies make policy incrementally; they often look like the "organized anarchies" that are depicted in the garbage can model of organizational choice.

Agencies may seek to expand the bounds of rationality with risk and economic analyses, but it is a battle. Whether it is a battle worth fighting is an issue we return to in the last chapter.

CHAPTER FIVE

Problems

I have referred to problems under various labels—global warming and ozone depletion, erosion and deforestation, lead and asbestos, health and ecological effects, wetlands loss, contaminated groundwater and radon, to name a few. These are common categories that most of us—policy makers, politicians, the lay public—use. If we look closely, we can see that these labels incorporate many different approaches to defining environmental problems. Take global warming. Is warming itself the problem, or is the problem the use of the fossil fuels and CFCs that generate the greenhouse gases that cause the warming? Or is the problem the sea level rise, changes in forests, or losses of ecosystems that may result from the warming? Another example is wetlands loss. Should the problem be defined as the losses in wetlands acreage, the threats to the survival of species and ecosystems, or the conversions to agricultural and commercial use that cause these effects?

In an incremental policy system, like that of the United States, problems are the raw materials of policy. Policy making is problem centered. The premise of our constitutional system is that governmental power exists to be contained, not promoted. Power is dispersed among institutions in a system of checks and balances and separation of powers. As a result, policy making is reactive, pragmatic, and incremental, rather than anticipatory, visionary, and comprehensive.[1] A condition must be seen almost as a crisis to warrant a response from policy makers. So the study of problems—how they are defined, organized, compared, and ranked—is vital to understanding how policy is made. If definitions of problems are fragmented and reactive, then our strategies will be as well.

Take the problem of urban air quality. In the Clean Air Act, policy makers viewed it as an environmental issue and responded with (among other things) technology controls on cars. If they had looked at urban air quality as an energy or transportation issue, they might have responded instead with transportation planning, industrial siting, and incentives for the efficient use of energy. Another example is water quality. Having accepted the notion that water pollution comes mostly from large, industrial sources, policy makers responded with technology-based controls on those sources. Yet water quality also is impaired by various nonpoint sources, such as agricultural runoff and storm water. A primarily nonpoint strategy would have consisted of other kinds of tools, such as land-use controls, management practices, and financial incentives.

Problems are conditions that are significant because they have undesirable effects. They are not objective phenomena but subjective constructs that acquire significance for any of several reasons—all related to their undesirable consequences. What is defined as a problem by one person may not be by others; what is viewed as a problem today may not be viewed as a problem tomorrow. Many people are convinced that emissions of greenhouse gases cause global warming. They see global warming as a problem that will have serious, irreversible consequences and that deserves a policy response. Others are skeptical. They oppose action until the scientific evidence is more definitive. To one person, an excess cancer risk of one in a million may not be a problem; another may view it as an excessive risk that deserves government attention.

This chapter considers environmental problems from three perspectives. First, I look broadly at their characteristics. Second, I discuss how to define and organize problems, with four profiles of some current problems. Third, I discuss agenda setting—deciding what problems deserve the attention and in what order—as well as risk-based environmental planning.

SOME CHARACTERISTICS OF ENVIRONMENTAL PROBLEMS

Let us begin with four propositions about environmental problems: they are embedded firmly in the public mind; they are interrelated and interdependent; they affect societies at several stages of growth; and they transcend national boundaries.

(1) *They are embedded firmly in the public mind.* Most observers would agree that environmental problems did not even begin to compete for a visible place on the national agenda until the late 1960s and early 1970s. As a result of the social activism of the day, a critical evaluation of the effects of economic development, and the growing recognition that natural resources and public health were at risk, environmental values took hold of the public mind. Focusing events, like the oil spill in Santa Barbara in 1969, reinforced this. Congress responded with the NEPA in 1969 and other environmental laws during the 1970s. Environmental groups emerged as advocates and guardians of the agenda. The environmental movement was born.

Yet the hold of environmental issues on the public mind was still tenuous through the 1970s. People were painfully aware of the tradeoffs between the nation's environmental and economic goals. The economic problems of the 1970s made the choices seem stark. The energy crisis further tested the public's commitment to the environment. Politically, the Reagan years posed a challenge to existing levels of political support for the environment.[2] The Reagan administration set out to reduce or eliminate regulatory programs as well as funding levels for EPA and other agencies. Public support for environmental programs was sorely tested, but it proved resilient.

A look at public opinion documents these trends. Here two concepts used by political scientists—those of strength and salience of public opinion—are important. Strength describes the degree to which people see an issue as important and want to do something about it. Polls measure the strength of an opinion on an issue by asking how serious it is or presenting people with choices between social objectives, such as the economy, on the one hand, and environmental quality, on the other. Salience describes "the amount of immediate personal interest people have in an issue."[3] Polls measure salience by asking people which one or two problems are the most important ones facing the nation.

The strength of public support for environmental programs was fairly strong in the early 1970s and has grown since then.[4] For example, the percentage of the U.S. public saying that we were spending too little to solve environmental problems grew from 45 percent in 1973 to 62 percent in 1988. The environment tied with education as the issue with the largest increase in public support during this period; at the same time, public support for energy and transportation declined. Similarly, the percentage agreeing that environmental protection is so important that standards could not be set too high, regardless of cost, grew steadily from 45 percent in 1981 to 65 percent in 1988. After the *Exxon Valdez* oil spill in 1989, it jumped to 80 percent—testimony to the role of focusing events in shaping public perception of problems.[5]

Even more significant is the trend in salience. Until recently, the environment was low in salience, overshadowed by such issues as the economy, inflation, war, and drugs. When asked what one or two issues they believed were the most important ones facing the nation, only 2 percent of the public named the environment as recently as 1987. By late in 1989, this had grown to 6 percent; after the *Exxon Valdez* oil spill, it had grown to 16 percent. Surveys from the early 1990s confirm these trends in strength and salience.

(2) *They are dynamic and interdependent.* Environmental problems are as dynamic and interdependent as any that government must face. Changes in the source, severity, or scope of one problem affect others. Solutions to one problem create new ones. Old problems recede or even disappear, but new ones emerge in a steady stream. One problem may have no relation to others; some are so intertwined that dealing with one and not others is a recipe for failure.[6]

Consider a brief illustration. Chlorine is used to disinfect drinking water before it is distributed. Using chlorine to kill pathogens in drinking water was one of the most important public health innovations of this century. Yet chlorine interacts with the bacteria that remain in the water to form by-products that present chronic health risks of their own. Scientists are studying the health effects of such by-products of disinfection to determine whether changes in chlorination practices or shifts to other water treatment technologies are necessary. There is no doubt that the benefits of disinfection outweigh the risks. Yet policy makers have had to assess the effects of disinfection to determine whether it creates other risks to which they should respond.

In the 1980s, concerns about the health effects of asbestos led Congress to authorize a national program to remove it from schools and other public buildings. Asbestos is a threat when it is not intact as it is likely to be released into the air and absorbed into the lungs. In their zeal to remove asbestos, building owners often disturbed intact asbestos or hired unqualified contractors to remove it. This created problems where none had existed. Besides promoting asbestos removal when there are risks, EPA has educated the public not to disturb intact asbestos and create health risks where none had existed.

New problems emerge on the agenda regularly. Problems like radon, depletion of stratospheric ozone, global warming, or the effects of biotechnology were barely on the agenda as recently as 1980. Now all but the last rank high as likely sources of risk.

(3) *They affect societies at all stages of development.* Given our experience, many of us think of environmental problems as products of the industrial age and as limited to the economically advanced societies. Yet beyond a very few small populations around the world, environmental degradation is present at all stages of development.[7] In fact, in preindustrial or developing societies, environmental problems often are more compelling than they are in the developed world.

For this discussion, I will consider nations in four groups, according to their stage of development and success at dealing with environmental problems.[8] First are the advanced industrial or postindustrial nations (the United States, Great Britain, Western Europe, and Japan). They industrialized over the last century (Britain or the United States), or more intensively since World War II (Japan), and coped with the consequences of their economic development. Their problems are familiar: air and water pollution; exposures to asbestos, lead, and other toxic chemicals; contamination of groundwater and other resources; workplace exposures to chemicals. The problems are difficult but manageable, given resources or changes in lifestyles—at least in comparison to those of other nations.

A second group has industrialized, mostly in this century, but has made little effort to cope with the consequences. Nations of the former Soviet Union and Eastern Europe are examples. The Iron Curtain lifted in the late 1980s to reveal a bleak landscape of environmental devastation. Upper Silesia, an industrial region in southwestern Poland, "is arguably the most polluted place in the world."[9] Life expectancy is falling, rates of leukemia and other diseases are rising, many children must be treated continually for chronic illness caused by the environmental conditions, and local produce is too contaminated with heavy metals to be eaten. The list could go on. The Aral Sea basin of Soviet central Asia and Kazakhstan is one of the most seriously damaged ecological zones in the world. The rapid withdrawal by farmers of irrigation water from rivers feeding into the Aral Sea has caused a drastic drop in its water levels; in three decades, its surface area shrank by over 40 percent and its volume decreased by more than 60 percent, leading to a variety of serious ecological and health problems in the area.[10]

The experiences of Eastern Europe and the former Soviet Union teach us that whatever we may think of the costs of environmental protection, the costs of neglect are far greater. Stressing growth and production to the exclusion of other values has caused nearly incalculable losses, in economic, health, and environmental terms. These losses take a tangible form; the Polish government estimated that the health effects from environmental neglect place a drag on its GNP of about 15 percent.[11] The environment may be salvageable in these countries, over time, but environmental restoration and protection will be difficult and expensive and will require outside help.

A third group is still in the midst of intensive industrial development. Examples are the dynamic economies of Taiwan, South Korea, and Mexico. Pollution of many kinds is a serious problem. Rapid growth, urban congestion, the high use of fossil fuels, often limited land and other resources for waste disposal—all put stress on the environment. Like Eastern Europe, these nations focused on economic goals, often to the exclusion of environmental values. They differ, though, in having suffered perhaps less harm so far and in having greater financial resources. Their economic success and promise at least provide them with the financial basis for dealing with their environmental problems.

Another group of nations, mostly in sub-Saharan Africa (Ethiopia) but also in Asia (Bangladesh), face basic problems of sustainability. They cannot feed their populations, because their resource base for producing food and sustaining life is declining. Soil erosion, encroaching deserts, and loss of forests proceed along with social and political breakdown to threaten the survival of their populations. It is in these nations that the prospects for coping with environmental problems—and most any other kinds of problems—are bleak.[12]

(4) *They transcend national boundaries.* A clear trend in the evolution of environmental problems is that they transcend national boundaries. This poses a dilemma for policy makers, who usually must deal with problems in the context of national laws and institutions.[13] Now problems must be solved on an international and global scale, requiring the institutional changes discussed in chapter 3.

We can identify three levels of transboundary problems.[14] First are "one-directional" problems, wherein the actions of one nation create environmental risks in others. In many of these cases, pollution moves through natural media. Prime examples are the pollution of the Rhine and acid rain. In the Rhine, industrial sources in several nations discharge into a shared waterway and impair water quality downstream. In the case of acid rain, the long-range movement of utility and industrial emissions affects water quality, forests, and other resources in nations that suffer the risks but enjoy none of the economic benefits of the uncontrolled emissions. In other cases, problems result from trade or export practices, such as the export of pesticides and hazardous wastes. The distribution of risks and benefits in one-directional problems makes them difficult to solve, at least from the receiving nation's point of view.

Second are regional problems, in which two or more nations both receive and transport pollution. Pollution in the Baltic Sea or the Mediterranean Sea is an example. With reciprocal problems, each country recognizes a joint interest in the careful use of a shared resource. Because there are incentives for reciprocity, the more successful efforts at cross-national cooperation fall into this category. The lines between "regional" and "global" problems are not always clear. Some problems, such as the destruction of tropical forests, are mostly regional in their causes and immediate effects but can have global implications.

Third are global problems, in which the actions of several nations combine to cause health and ecological effects that are broadly distributed, with no direct relationship between the sources of pollution and the effects associated with them. Risks and benefits are spread unequally, across regions, among nations at various stages of development, and across generations. The actions of one nation, even of regional groupings of nations, will have only marginal effects on these problems. As a result, efforts to understand and respond to them must be global in scope.

Changes in climate are the most visible—and in the long run, the most significant—of these global problems. Depletion of the stratospheric ozone layer due to the effects of CFCs, halons, and other gases will have many consequences, such as increased incidences of skin cancer and major impacts on ecosystems. An even more profound problem is global warming, also known as the "greenhouse effect." Many greenhouse gases occur naturally in the atmosphere. They act as a thermal blanket that allows radiation from the sun to enter the atmosphere and strike the earth and then absorbs the infrared radiation that is emitted from the earth's surface. As the concentrations of greenhouse gases rise from human activity, so will atmospheric temperatures. Since 1880, atmospheric CO_2 levels have risen 25 percent and may have increased temperatures by 0.5 to 1.1°F. The National Academy of Sciences concluded in 1991 that there is a "reasonable chance" of a doubling of preindustrial CO_2 levels by the middle of the next century and that this "could ultimately increase the average global temperature" by between 1.8 and 9°F.[15]

DEFINING ENVIRONMENTAL PROBLEMS

This section considers four problems that currently have the attention of policy makers. They are but a sampling of what is on the policy agenda, but they illustrate the variety in environmental problems and the ways in which policy makers compare, evaluate, and rank them. Before moving to the case studies of the problems themselves, however, we will look at how environmental problems may be defined and organized as a basis for action by policy makers.

WAYS OF ORGANIZING ENVIRONMENTAL PROBLEMS

Later in this chapter I examine EPA's *Unfinished Business* report. Its purpose was to compare and rank environmental problems according to the health and ecological risks they posed. It broke the universe of potential environmental problems into a list of thirty-one specific problems.[16] Varied and often overlapping principles were used to define the problems. Some were defined as "contaminants" (radon or air pollutants), others as "sources" (water discharges from point sources or hazardous waste sites), and still others as an "affected resource" (groundwater or wetlands) or a "medium of exposure" (drinking water or the ambient air).

Four years after *Unfinished Business* was published, EPA's Science Advisory Board proposed its own approach to defining problems, based on the effects of environmental "stress." SAB defined problems based on five categories: the *activities* and *contaminants* that create environmental stress; the *medium* or pathway through which those contaminants are transmitted; the *receptors* that are affected by the stress; and the *long-term effects* of environmental stress.[17] Table 8 gives a summary.

TABLE 8 DEFINING ENVIRONMENTAL PROBLEMS: ONE APPROACH

Defined on the Basis of	*Examples*	*Comments*
Activity causing the stress	Energy production Agriculture Transportation	Gives a basis for an integrated policy or industrial sector approach to pollution control
Agent or pollutant causing stress	Lead Radon Asbestos Ozone	Usually used to identify high-priority threats to human health across environmental media
Medium or route of exposure	Drinking water Ambient air Pesticides on food Surface water	The approach most often used in U.S. pollution control laws and programs
Target or receptor affected	Groundwater Wetlands Coastal resources	The basis for many water and ecological protection programs and for geographic approaches (e.g., the Great Lakes)
Effect of the stress	Losses in biodiversity Global climate change	Defines broad-scale, global problems, often with irreversible effects

SOURCE: Adapted from *Reducing Risk: Setting Priorities and Strategies for Environmental Protection* (Washington, D.C.: U.S. Environmental Protection Agency, September 1990).

Activities One approach is to define problems based on the activities that cause environmental stress. Energy and agriculture account for a large proportion of these stresses. Energy production and use create risks as diverse as oil spills, acid rain from burning fossil fuels, disposal of radioactive waste from nuclear power plants, or carbon monoxide from vehicle emissions. Agriculture is a source of many stresses, among them, application of pesticides, cultivation of highly erodible land, depletion of nutrients in soil, and the leaching of nitrogen fertilizers into groundwater.[18] Mining, chemical production, and transportation are other activities that create environmental stress. Defining problems in this way, that is, according to the activities that create stress, allows policy makers to design integrated strategies that focus on policy sectors or industries.

Contaminants Another way to define problems is to focus on the contaminants that cause the stress. Such contaminants as lead, asbestos, ozone, and radon are found commonly in the environment and present documented health risks. The focus for policy is the contaminant that causes harm throughout its life cycle—through production, processing, distribution, use, and disposal. Many cross-media approaches to solving environmental problems are organized on the basis of the contaminant, such as the lead strategy that is discussed in chapter 6.

Medium or Pathway of Exposure We can define problems based on the medium or pathway of exposure. How is it that humans or ecosystems are exposed to environmental contaminants? Humans are exposed when they ingest, inhale, or come into skin contact with substances. Contaminants in drinking water, pollutants in ambient air, and pesticide residues in food are common pathways of exposure. This is the most common way of organizing problems under current laws and programs. For the most part, the Clean Air Act, the Water Quality Act, and the Safe Drinking Water Act rely on media-based definitions, as does EPA's organization, the organization of most state agencies, and the oversight structure in Congress.

Receptors A fourth way to define problems is on the basis of the organisms, resources, or ecosystems that are affected by the environmental stress—whatever its cause. A basic distinction is between humans and ecological resources as receptors. Several very visible problems, for example, coastal or estuarine resources and groundwater, are defined on this basis. The attention given to such issues as groundwater and wetlands demonstrates how important resources (considered as a kind of receptor) are as a category of problems. This approach allows policy makers to focus on geographic areas that define a "receptor," such as the Great Lakes or Chesapeake Bay.

Long-Term Effects Another approach to defining a problem is by focusing on the long-term effects of environmental stress. Two examples are global climate change and losses in biodiversity. Climate change is the result of a range of human activities, most related to the use of fossil fuels, others to agricultural or industrial activity. Losses in biodiversity are similarly the product of a variety of activities—commercial development and agriculture, for example. By defining problems in this way, policy makers can get a much broader view of the relationships among systems in the environment, of the long-term implications of problems, and of the need for more comprehensive policy responses.

FOUR PROBLEMS ON THE NATIONAL POLICY AGENDA

Two of the problems discussed below are defined by contaminant (radon and lead), a third by the receptor (groundwater), and the fourth by the long-term effects of stress (losses in biodiversity). They are only four of a long list of problems, but they illustrate a useful mix of approaches, kinds of risk, and other factors. They can be compared in several ways: Do they pose health, ecological, or a combination of risks? What is the magnitude and severity of risk? Does the problem occur nationally, or is it limited to a geographic region? Is the problem attributable to natural causes or to human activity? Where does the problem fall on the policy agenda, and how recently has it been recognized by policy makers as a problem?

Radon and Lung Cancer Radon is a naturally occurring gas that is invisible and tasteless. It is produced by the radioactive decay of the element radium, itself a product of the decay of uranium or thorium, which are two elements found in low concentrations in rock and soil. Radon does not react chemically with most materials; it travels freely as a gas and moves easily through small spaces. Radon is moderately soluble in water and can be absorbed by water flowing through rock or sand. Exposure occurs from inhalation of the gas and, when radon is in solution with water, through ingestion or absorption through the skin.

The federal government considers radon to be one of the nation's most significant environmental health threats.[19] It is the second leading cause of lung cancer, after smoking, and has been studied extensively. Estimates of health effects are based on epidemiological studies of underground uranium miners in the United States, Canada, Sweden, and elsewhere. EPA estimates annual cancer deaths from radon exposure in the United States at about 14,000—a "central" or "most likely" estimate within a range of 7,000 to 30,000 deaths.[20] Because these estimates rely on studies of humans, in this case, thousands of miners, they are more reliable than data from animal studies of other substances. Still, critics argue that EPA's risk assessments are conservative. "The data, many of which are based on high exposures in dusty unventilated mines, have been extrapolated to low doses in relatively dust-free living rooms."[21] The critics argue that at low doses and in the much cleaner setting of people's houses (compared to a uranium mine), actual risks would be lower.

Health risks do not fall equally on all who are exposed. The most important factor that increases risk within the population is smoking. The relationship between smoking and radon exposure is multiplicative rather than additive; smokers are ten times more likely to develop lung cancer from given levels of radon exposure as are nonsmokers. Children also may be at special risk. They are more vulnerable to damage from radon, and exposure at an early age increases risks of lung cancer over their lifetimes. In its analysis, EPA assumed that the risks to children would be three times those to adults, as a way of building in a safety factor.

Although the potential risks of exposure to radon have been known for years, it has emerged relatively recently on the policy agenda. Risks to uranium miners had been documented for some time, beginning with studies from the early 1950s. Residential exposures to radiation drew official concern when high levels were discovered in homes made from materials that contained phosphate slag or tailings from uranium mills. Although scientists were aware that radon occurs naturally under homes and can build up under certain conditions, residential exposure from naturally occurring radon was not seen as a major problem until a 1984 incident involving a worker at a nuclear plant in eastern Pennsylvania. Each day when he left work, Stanley Watras activated alarms on the radiation monitors. One day, he decided to test the source of the alarms by turning around and passing the monitor *before* he entered the plant. Again, the alarm went off. When officials tested the air in his home, they found the highest radioactivity yet measured in any home in the world—with health effects equal to smoking 135 packs of cigarettes a day.[22]

Since then, radon has risen quickly on the agendas of agencies like EPA and the Public Health Service and, to some degree, of Congress. It has risen less on the public's. The low level of public concern has puzzled many health officials, who see in radon a greater threat than exists from many other problems that receive more attention. The reasons may lie in its characteristics as a source of risk. A quick look at the research on risk perception suggests why radon is the kind of problem that will not greatly alarm the public. Radon occurs naturally and is undetectable to the senses. It poses a statistical risk, in the form of the predictions of future cases of lung cancer, and its effects are not immediately observable. Exposures to radon in homes is in a sense voluntary (nobody is pumping it into houses, and people can act to greatly reduce levels of radon) and is not imposed on people by others. And finally, the effects of radon are discrete—individual cases of lung cancer that occur over time—and not catastrophic.

Contaminated Groundwater Half the people in the United States depend on groundwater as their source of drinking water.[23] In rural areas, the number is 97 percent. About 90 percent of the population in Florida, Hawaii, Mississippi, and New Mexico get drinking water from underground; in five other states, more than 75 percent do. Groundwater is also the source of 40 percent of the water used to irrigate cropland and about 25 percent of the water used by industry. It serves ecological needs as well; it is a source of water recharge for surface streams and helps to prevent the intrusion of salt water into water supplies in coastal areas.

Although most groundwater in this country retains its natural purity, there are areas where enough contamination has occurred to make the water unsafe. For example, the insecticide and nematicide Aldicarb was used on 24,000 acres of potato fields on Long Island between 1975 and 1979. In 1980, testing found Aldicarb levels that exceeded New York's drinking water standards in nearly one-third of the samples. Nitrates (compounds with a form of nitrogen) often are found in rural areas. A survey of rural water quality found detectable levels of nitrates in 57 percent of the rural domestic wells that were tested.[24] Other studies have found many contaminants—heavy metals like lead or cadmium and synthetic organics like benzene—in aquifers. In some areas, people use bottled water, treat tap water to remove contaminants, or install barriers to contain the movement of contaminants through an aquifer.

Groundwater can be contaminated in several ways. The most important are related to the disposal of waste. Home septic tanks that dispose of sewage, underground injection wells used to store oil and chemical wastes, impoundments for storing waste on the land surface, sanitary landfills and dumps that hold municipal wastes, and active or abandoned hazardous waste facilities all can affect groundwater. Leaks from underground tanks used to store petroleum and chemicals are another source. Agricultural activity is a major source of contamination, from the application of fertilizers and pesticides, from irrigation water that seeps into the ground, and from manure from livestock operations. Contamination also results from polluted surface waters, salt that is used to de-ice roads, storm water runoff, and the ground deposition of air pollutants.

Groundwater contamination poses both health and ecological risks, although public policy has focused on the first. Health effects depend on the contaminant. Many found in drinking water supplies are known or suspected carcinogens. Several are linked to other acute and chronic effects. Nitrates, for example, may be linked to nervous disorders, birth defects, and methemoglobinemia (blue baby syndrome). The ecological effects of groundwater contaminants have been less prominent on the agenda but could be significant. The close relationship between groundwater and surface water suggests that ecological damage aboveground is linked to problems below ground. In addition, groundwater has an "existence value" that should enter into any evaluation of risks. Even when people are unlikely to use an aquifer for drinking water, they attach value to its existence in an uncontaminated state.

Groundwater contamination occupies a prominent place on the nation's environmental agenda. Much of the policy initiative on groundwater has come from the state level. In local areas where actual evidence of contamination has raised concern about health risks or made water unfit for drinking, it is a highly salient and visible issue. At the federal level, the concern about groundwater manifests itself mainly in the RCRA and Superfund programs, although EPA has laid out a broader, cross-media policy.[25]

Lead Lead is an element and the most abundant heavy metal in the earth's crust.[26] Its versatility and many functions have made it a commonly used material since ancient times, and today it is one of the most pervasive toxic pollutants. Lead ranks among the oldest environmental threats to health; some historians attribute the decline of the Roman Empire to the neurological effects of ingesting lead that leached from the lining of drinking vessels. Risks from lead are mostly to health, but there is evidence of ecological risk as well. Lead has been studied extensively, and its health risks are well documented. Children are especially vulnerable, due to behavioral factors (playing in soil, eating paint chips) that increase their exposure and physiological factors that increase the rates at which they absorb and retain lead in their systems. Exposure to high levels can cause lead poisoning and death, but there are treatments for removing lead from the circulatory system, and deaths from acute poisoning are rare.[27]

A greater risk is damage from prolonged exposure at lower levels. Lead accumulates in the central nervous system, and it can seriously impair the mental and behavioral development of children. The mean

IQs of children with elevated levels of lead in their blood have been shown to be up to five points lower than the IQs of children with low levels. Even at the lower lead levels found in many drinking water systems, the average IQ could be reduced by a point. This constitutes a large effect when it is spread across the hundreds of thousands of children who may be exposed at such levels. Other effects on children are neurological (hyperactivity, seizures, hearing impairments), hematological (anemia due to destruction of red blood cells), and renal (kidney malfunction). Risks to adults are cardiovascular, reproductive, and osteopathic. Most significant are the risks of hypertension and possibly osteoporosis, miscarriages, and testicular disorders. There is evidence of a link between lead and several cancers, but it is inconclusive. EPA classifies lead as a "suspected" carcinogen.

About half of global lead production each year comes from the mining of virgin lead (some 3.4 million tons in 1988); the rest is from the secondary production of recycled lead. Global consumption of lead rose rapidly in the 1960s, then leveled off. Nearly 60 percent of the demand for lead in the United States and Western Europe is for lead acid batteries. Cable sheathing accounts for another 10 percent. Other uses are as fuel additives and in water supply piping and solder, metal alloys, oxides, and chemicals. The use of lead as a fuel additive and for water piping declined as various regulatory limits took effect, though there still are significant exposures from drinking water. Much of the health risk today comes from lead that has remained in the environment from past uses. Lead is not used as a pigment in paint today, for example, but children inhale or ingest dust and chips from paint in older houses. Similarly, lead emissions from smelters or battery cracking operations have been eliminated or greatly reduced, but lead has built up in soil where children now play.

Lead in paint and soil poses serious risks for children living in inner cities, where elevated blood levels are most common.[28] Many urban areas are seen by environmental and public health officials as "hot spots" of lead exposure. Lead remains high on the nation's environmental and public health agendas. In 1992, Congress passed the Residential Lead-based Paint Hazard Reduction Act, which directed EPA and other federal agencies to take several actions to reduce risks from lead paint, especially to children in urban areas.

Losses in Biodiversity "Biodiversity" is a term for describing the variety of life at all levels. It has always been on the environmental policy agenda in some form, but it has become more influential recently as a category for defining and organizing responses to many specific problems. Biodiversity is an implied goal of many government programs that are designed to protect wildlife, natural habitat, and endangered species, but it is far more inclusive than the concepts underlying most of these programs. For example, if we frame the problem of "endangered species" as a loss of biodiversity rather than just the loss of individual species, we are more likely to recognize the value in all forms of life, at several levels of ecological organization, not just those that are immediately threatened or especially valued. Biodiversity is a concept that captures an array of ecological values and effects.

Think of biodiversity at three levels: genetic diversity within species, both among individuals in a population and among different populations; species diversity, or the numbers and frequencies of species in a region; and ecosystem diversity, or variety in the communities of organisms in a physical setting.[29]

The threats to biodiversity are numerous. They are "primarily the product of two trends: the exponential growth of the human population and the increasing per capita consumption of the Earth's resources to provide for human needs."[30] Alterations in physical habitat are one threat. Each year, for example, some 300,000 acres of wetlands are converted to cropland or commercial uses. Even partial conversions of wetlands to other uses (damming rivers or building roads) fragment and damage ecosystems. Such alterations in habitat pose a major threat to biodiversity.

A second category of threats is pollution and pesticides. After decades of focusing on health risks, policy makers have begun to appreciate the ecological harm that many common pollutants may present. Among air pollutants, ozone and acid rain appear to be especially damaging. Ozone damages coniferous and deciduous forests. For example, pine trees not only grow more slowly if exposed to ozone but are also more susceptible to pest damage. Aquatic ecosystems are hurt by the buildup of nutrients like nitrogen and phosphorus. Runoff from agricultural lands and from wastewater treatment plants are two major sources of nutrients. In the process of eutrophication, excessive levels of nutrients in fresh water stimulate the growth of algae, whose later death and decay remove oxygen that fish and other life need to survive. Pesticide application on cropland and even on lawns poses threats to many organisms and ecosystems. Some chemicals are so toxic or occur at such high levels that effects are immediate; pesticides that kill songbirds are an example. Others build up at a slower rate and bioaccumulate (build up over time in biological systems).

A third category of threats is climate change, which alters the biosphere and may dramatically affect diversity.[31] Warming would alter vegetation patterns, especially at the center of landmasses and at higher latitudes. More significant than the warming itself would be its effects on other aspects of the global climate—patterns of rainfall and winds, movements of air masses, patterns of ocean circulation, and heat distribution on the earth's surface, among others. Effects on hydrological cycles would be important. Warming would melt snow earlier in the year; make many ecosystems drier; and make some large land areas wetter, others drier. Increased ultraviolet radiation due to ozone depletion poses additional threats. There also are direct effects from increases in atmospheric carbon dioxide levels, aside from the warming phenomenon itself, primarily affecting plant physiology.

Biodiversity achieved some standing at EPA as a result of a 1991 report by the agency's Science Advisory Board, which urged EPA to consider biodiversity as a priority in its programs. Losses in biodiversity and other long-term effects of environmental stress figured prominently in the report, "because of the pervasive extent of these environmental stresses and the diversity of resultant impacts on ecological systems at species, community, and process levels."[32] Biodiversity is not likely to dominate the political agenda, where attention focuses on actions that affect ecological quality (conversion of a wetland, loss of habitat for migratory birds, effects on wildlife of a specific pesticide) rather than on the long-term, systematic effects of these actions. Yet the concept of biodiversity may prove influential in shaping the strategies of government agencies toward the many sources of ecological risk.

COMPARING THE PROBLEMS

Radon presents a health risk—the risk of lung cancer. Lead poses several noncancer health risks, mostly in its developmental effects on children. Contaminated groundwater may pose serious health risks of specific kinds, from chronic cancer risks from toxics to the range of effects associated with nitrates. Contaminated groundwater tends to be a localized problem, but it has a place on the national policy agenda and was the impetus for the RCRA and Superfund programs. It threatens surface water quality as well, with potentially serious ecological effects. Loss in biodiversity is the broadest-scale problem of all, with complex and perhaps irreversible effects.

Each of these is a major environmental issue, but together they represent just a few among many issues that face policy makers today. Now we turn to another question: Given all the problems that demand attention from policy makers, how do we decide which of them are important enough to warrant a policy response from government, and in what order should government respond?

SETTING THE POLICY AGENDA

The challenge in environmental policy is the classic political and administrative need to set priorities. Political institutions set priorities at many levels. Think of a hierarchy of choices: defense matched against domestic needs, educational against environmental goals, the benefits of resource exploitation against those of preservation, and so on, through budgetary, legislative, regulatory, and other processes. In environmental programs, the same kinds of choices and trade-offs must occur. We cannot deal with all the problems that warrant attention, so we make choices.

Environmental policy is replete with choices. How do we compare an ecological risk like the loss of a wetland with a health risk like the increased incidence of lung cancer near a chemical plant? Does the contamination of an aquifer in a lightly populated area warrant as much attention as damage to an estuary, when that threatens commercial fishing? Should we give more attention to the long-term risks of formaldehyde in building materials than to the unlikely but potentially catastrophic effects of emergency chemical releases? Of the problems profiled above, which is more worthy of a place on the agenda: Lead and its effects on children's development? Groundwater pollution, which takes years to attenuate naturally or lots of money to restore through treatment? The lung cancer risks linked to radon exposure? Or long-term, widespread, irreversible losses in biodiversity?

Social and political processes for selecting among problems and ranking them for a response are the subject of agenda setting. Its focus is the policy agenda: "the list of subjects or problems to which government officials, and people outside of government closely associated with those officials, are paying some serious attention at any given time."[33] Agenda setting is an important stage in the policy process. To get attention, a problem must earn a place on the agenda; to get a response from policy makers, it must have priority over the other problems that are competing for attention. Because there are more problems than there is room on the agenda, problems compete for space. Hazardous waste or global warming may rise quickly to prominence; biotechnology or radon may get official attention but may stay on the edge of the public agenda.[34]

Agenda setting is the process by which the large number of problems government *could* address is narrowed to the small number of problems government *will* address.[35] It is a complex process in which people, events, and timing all play roles. Demographic and economic trends are key influences. Events are also important. The accident at Three-Mile Island in 1979 moved nuclear safety up on the agenda. It was a focusing event that gave a concrete expression to public fears about nuclear power. The efforts of "policy entrepreneurs" also influence the agenda. One reason that environmental laws fared so well in the late 1960s and early 1970s was Senator Muskie's interest in carving out a leadership position to promote his presidential campaign.

Events clearly shape the agenda, especially focusing events. The *Exxon Valdez* oil spill had dramatic effects on the strength and salience of public opinion. It had tangible institutional and policy effects also, including the Oil Pollution Act of 1990 and a general strengthening of relations among the federal and state agencies that were responsible for responding to oil spills. The accident at the Union Carbide plant in Bhopal, India, in December 1984 and the deaths of some three thousand people focused public and official attention on emergency releases of toxics. It led Congress to pass a community-right-to-know law in the reauthorized Superfund law of 1986 and to create an accident investigation board.

FOUR CASES IN AGENDA SETTING

A few examples show how varied are the factors that move issues up and down on the agenda. Here is a look at hazardous waste, the pesticide Alar, medical waste, and global warming, at how each took and in some cases lost a place on the policy agenda.

Love Canal and Hazardous Waste Hazardous waste emerged as a major environmental issue in the 1980s. The Superfund program now constitutes a very large category of EPA spending. Overall, the costs of complying with waste programs have grown steadily and are projected to rise to more than $46 billion annually by 2000, one-third of total U.S. environmental spending. Yet in two reports that will be discussed below, EPA's *Unfinished Business* and the Science Advisory Board's *Reducing Risk*, hazardous waste ranked medium to low as an ecological and health risk. So in EPA's own statement of its priorities, as opposed to what political and legal pressures have made them, waste ranked in the middle in risk.

As recently as 1976, when the RCRA was passed, few people would have predicted the rapid rise of hazardous waste on the agenda. One book noted that the "environmental revolution of the 1960s had left solid waste largely untouched. The problem was viewed as one of materials handling, routing garbage trucks, and bulldozing landfills . . . as the province of local governments."[36] EPA's position was consistent with this view. Passage of the RCRA in 1976 changed this to some degree, by carving out a federal role in waste management that had not existed previously. Yet EPA had not played actively in the RCRA's passage and had invested little in resources and political commitment in waste programs. Even after the RCRA passed, it was something of a backwater in EPA. Congress showed only sporadic interest. Environmental groups gave waste issues a low priority; they preferred to focus on water, air, and the new toxic substances law (the TSCA), where expectations were high.

By the 1980s, the place of hazardous waste on the national policy agenda had changed. The focusing event came in August 1978, when *The New York Times* ran a story on an abandoned waste site near Niagara Falls, known as Love Canal. *Time* and *Newsweek* followed with major articles, as people were treated to stories of waste oozing into basements, of odors and possible incidents of cancer, and of other problems. The effect of all this publicity was to focus public attention on hazardous waste. "Under the prodding of intense media coverage, the public began to view hazardous waste as a serious national problem."[37]

NRDC and the Alar Controversy With many sources of chronic health risk, focusing events are unlikely. Chronic risks are the product of years of exposure; the links between exposures and effects are not obvious. Yet events may draw publicity to chronic risks, especially if they threaten populations of special concern.

The controversy over the pesticide Alar in the winter and spring of

1989 illustrates this effect. Artificial chemicals in food have long been a public concern, from red dye to saccharin. Government decisions about food additives or chemical residues are usually based on animal testing. When human effects are clearly documented, the regulatory decision may be easy. The controversy turns on the hard cases, those in which the evidence on chronic risks, usually cancer, is drawn from animal studies. Alar had been suspect for some time. EPA had nearly canceled its registration as a growth regulator on apples in the past. In 1988, EPA was waiting for the results of additional testing that could justify cancellation. At the same time, the Natural Resources Defense Council publicized the results of tests that showed higher risks from cancer than EPA's own testing had shown.[38]

NRDC skillfully used the media to promote its case against Alar. It stressed special risks to children, who consume large amounts of processed apples in juices and applesauce. NRDC painted a picture of millions of American children drinking and eating products laced with a cancer-causing chemical. It hired public relations firms to promote the story and help place it on "60 Minutes" and "Donahue." NRDC enlisted the actress Meryl Streep to make a commercial and to testify before Congress on Alar's hazards. Stores pulled apple juice from shelves; schools dropped apple products from their menus. Consumption of apples and apple products fell quickly, until the manufacturer took Alar off the market.

NRDC had turned a chronic health risk into a focusing event that pushed food safety up on the national policy agenda. Alar's health effects could not be demonstrated clearly, but NRDC created the perception of a clear health risk. EPA canceled Alar's registration (something it probably would have done in any event), but it was almost unnecessary at this point because of the impact of negative publicity. Alar rose quickly on the agenda and fell just as quickly off the market.

Medical Waste and Despoiled Beaches One of the biggest environmental stories of 1988 was a series of beach closings along the Atlantic Coast, especially in New York and New Jersey. Among the debris found on beaches were several kinds of medical waste, such as blood bags, syringes, hypodermic needles, and surgical wastes. Officials closed beaches for days at a time, and several resorts suffered large financial losses. There was broad media coverage.[39]

These events had an immediate effect. State officials announced new enforcement efforts, and several states passed or considered legislation.

EPA formed the Medical Waste Task Force to coordinate the work of several offices. Congress held hearings that led to the passage of the Medical Waste Tracking Act of 1988 by the end of the year. EPA issued rules implementing the act by March 1989. It was a powerful display of responsive government; a threat appeared, and Congress and EPA moved swiftly to deal with it.[40]

The catch was that the response was only marginally related to the events that had triggered it. Analysis of the beach closings showed that the bulk of them were the result not of hypodermic needles or blood bags strewn on beaches but high levels of fecal coliform from sewage that had washed up on shore.[41] In New Jersey, over 80 percent of beach closings were due to contamination from fecal coliform. "The actual composition of the washups contained 'a very small percentage'—about 10 percent or less—of 'medical-related' waste."[42] The problem was not poor waste handling but direct discharges of municipal sewage into coastal waters as a result of combined sewer overflows (CSOs). In periods of heavy rainfall, capacities of many municipal sewer systems are overwhelmed and raw sewage is discharged directly with storm water wastes. Whether this sewage ends up on beaches depends on weather and tidal patterns more than anything else. To avoid such overflows, New York and other (mostly eastern) cities will have to spend billions to separate municipal waste sewers from storm water sewers.

The waste handling and treatment provisions of the 1988 act were by no means bad law or a step backward, although there were complaints about the costs of the full program. It seems clear, though, that the law was not the solution to the events that had brought the problem of medical waste up on the agenda. If beach closings have not occurred at the same rate since then, it is most likely because of changes in tidal patterns or reductions in discharges of untreated sewage into coastal waters, not the effects of the medical waste tracking law.

Uncertain Science and Global Climate Change The rise of global climate change on national and international policy agendas during the 1980s was swift and impressive. What had been on the margins of the agenda a decade before moved to the center of the environmental debate. Climate change demonstrates some interesting dynamics of agenda setting with its confluence of science, media attention, political opportunity—and weather.

Global climate change consists of long-term trends whose effects may not be felt for decades but whose causes exist now. It is a problem that has been identified almost entirely on the basis of scientific prediction, not observable fact. We see garbage on beaches, monitor lead levels in drinking water, observe wetlands being converted into shopping centers, feel the effects of ground-level ozone on a hot summer day. But we cannot immediately see or feel the effects of climate change or ozone depletion. We rely on scientific analysis and prediction to define the existence, magnitude, and effects of the problems.[43]

Knowledge of the relationship between human activity and climate moved slowly until recently. Calculations by a Swedish scientist in the 1890s first linked concentrations of CO_2 with surface temperatures. Major advances in scientific understanding took more than half a century: the association between fossil fuels and CO_2 concentrations was established in the 1920s; industrial production of CO_2 and temperatures were connected more directly in the 1930s; the first CO_2 monitoring station began operation in Hawaii in 1958. The pace of knowledge grew rapidly after that, with the advent of computer models and research linking other gases (like methane or CFCs) to warming and ozone depletion. Despite the advances, uncertainties still limit our knowledge of climate change and the effects of greenhouse gases. Many scientists argue that the predictions fail to account for the negative feedbacks that will offset warming, among them heat absorption by oceans, greater cloud cover, or CO_2 enrichment that could encourage plant growth.

Unlike some issues, such as genetic engineering, climate change made the leap from the scientific to the political agenda. Helen Ingram and Carole Mintzer note that "there has been nothing slow and incremental about media and political attention to global warming."[44] The hot summer of 1988 was critical in this rise to prominence. From 1987 to 1988, the number of articles on climate that appeared in five major newspapers increased nearly tenfold, to over one hundred in 1988 compared to only a few at the start of the decade. Part of the increase was a result of the advocacy of scientists—Roger Revelle of the Scripps Institute of Oceanography, James Hanson of the Goddard Institute for Space Studies, and Michael Oppenheimer of the Environmental Defense Fund. But what drew media attention was an unusually hot and dry summer. Though scientific evidence of the relation between seasonal weather and long-term climate is thin, heat and drought made the problem immediate and relevant. Advocates of action on global warming effectively used events—short-term changes in weather—to move an issue up on the agenda.

Comparing the Four Cases in Agenda Setting These cases illustrate how issues move onto and (sometimes) off the agenda. Waste rose rapidly in the late 1970s and early 1980s to take a place with air and water pollution as a major issue. The dismal stories about Love Canal and many other sites offered people a visible sign of the effects of past neglect and lapses in current environmental laws. Hazardous waste became a ticking bomb in the 1980s. Waste problems tended to be localized, but with the stories of risks borne by residents of areas near waste sites raising much broader concerns, hazardous waste became a national issue. Many people wonder if the waste problem warranted a national response on the scale of Superfund and the RCRA.[45] But once established, programs have momentum; interest groups, legislators, and courts resist efforts to cut them back. Hazardous waste acquired a constituency.

Alar illustrates a different phenomenon. Pesticide residues on food are only one of many dietary exposures that concern us. There is no question that given the large number of exposures to pesticide residues in foods, they *should* concern us. Even small incremental risks are important if the exposures occur across the entire population of 250 million. At a one in a million (1 x 10^{6}) risk—one many would consider de minimus—there still could be 250 excess cases each year from residues in widely consumed foods. This is high compared to other environmental cancer risks.

Yet the question remains: Did Alar deserve to be canceled? Probably yes, at least when the evidence on Alar is compared to other pesticides. EPA would almost certainly have canceled Alar at some point, but it felt compelled to do more analysis before trying to make the legal case that would have been necessary under the FIFRA. There were costs to the way Alar was canceled in the end: economic damage to apple growers, more public concern than was necessary, perhaps a temporary fall in apple consumption by children. Congressional interest in food safety increased for a time, but the Alar issue had little lasting impact, and food safety faded quickly from the agenda.[46]

Medical waste presents perhaps the most curious case of all. It would be hard to argue that its improper disposal did not deserve a high level of public concern or official attention on the agenda. If it had not been for the wash-ups and beach closings, medical waste would have remained a minor issue. But events can push buttons in the public mind, and stories of hypodermic needles, tissue remains, and vials of blood on resort beaches pushed some. Add to this the growing visibility of AIDS in the media; surely the concern about medical debris was linked to awareness about AIDS. And so we have in medical waste a mismatch between a cause of alarm and a policy response. The Medical Waste Tracking Act may be worth having, but it will have little effect on whether debris washes up on beaches in the future. Until there are more beach closings or similar events, medical waste probably will stay off the agenda.

Hazardous waste moved up on the agenda and stayed there. Alar came and went, although the potential for a chemical of the week to take a visible place on the agenda remains. Medical waste enjoyed a short half-life but precipitated an immediate policy response. Global warming differs in some ways from each of these cases, because there is no visible manifestation of a problem, no waste in basements or syringes in the sand. As with the case of Alar, the conclusion that a problem exists depends on scientific assessments and predictions and is not apparent to the senses. Like studies of the health effects of pesticide residues on foods, these scientific analyses are controversial; no two experts necessarily will agree. Results of analyses vary, depending on the choice of models and assumptions. But issues like global warming move up on the agenda when marginally related events (heat and drought) draw attention to them and the salience of environmental issues is high, as it was in the late 1980s.

These are only four of many stories we could tell about how issues appear on and disappear from the agenda. Problems rise and fall depending on a variety of factors and events. Predicting what will come up next is even more difficult than explaining how what is on the agenda now got there. Will the next priority be the possible risks from electromagnetic fields? A "chemical of the week" that for some reason draws public attention? Greater focus on oil spills, if there should happen to be another accident on the scale of the *Exxon Valdez*? Or something that we cannot predict at all, like a major chemical accident, the release of a genetically engineered substance that destroys wildlife and crops, or the long-latent but irreversible health effects of a commonly used consumer product? And even for the problems we do know about, are we as a society allocating our scarce resources and talents wisely?

The final topic in this look at problems is an alternative approach to the task of setting the environmental policy agenda. Known as risk-based planning (or comparative risk), it is EPA's attempt to set priorities more rationally and comprehensively than policy makers have in the past and to stimulate debate about risk. It is an attempt to meet the challenge of setting priorities.

RISK-BASED PLANNING: A BETTER WAY TO SET PRIORITIES?

In January 1991, *The New York Times* ran an article entitled "E.P.A. Acts to Reshuffle Environment Priorities." It described testimony to the Senate Committee on Environment and Public Works by the EPA administrator urging a new approach to organizing and ranking environmental problems. The testimony drew on a report by EPA's Science Advisory Board that criticized how the nation typically had defined problems and set environmental priorities. One of the report's principal conclusions was that environmental problems ranked high by SAB "were often different from those that excited the public and stimulated Congressional action." For example, SAB ranked global warming and its effects among the high risks, and hazardous waste much lower. In contrast, public opinion surveys usually ranked hazardous waste high and global warming lower. The article described the report and the testimony as part of an EPA effort "to take a central role in shaping a coherent, long-term environmental policy based on scientific assessments of risk."[47]

This effort began long before the testimony was delivered. Four years earlier, EPA had issued *Unfinished Business: A Comparative Assessment of Environmental Problems*.[48] This report was a first attempt to evaluate problems differently than EPA, Congress, and nearly everyone else had in the past. After this report came several regional and state comparative risk projects, each of which built on the approach in *Unfinished Business*.[49]

The Unfinished Business *Report* Under Administrator Lee Thomas, EPA had formed four work groups to assess the state of knowledge about four kinds of environmental risk: cancer health, noncancer health, ecological risks, and a category called welfare risks, which included economic damages (such as the effects of air pollution on crops or the losses to commercial fishing from water pollution). The work groups assembled the available data on each category of risk and ranked an agreed upon list of thirty-one problems on the basis of these data. Problems seen to pose greater risk (based on the expert judgment of the work groups) ranked high; problems that the work groups decided posed lesser risks ranked in the middle or at the bottom of the evaluations.[50]

None of the problems ranked high in all four risk categories. Some of them did rank high or medium in three of the four categories, however. These included pesticide residues on food, "criteria" air pollutants (those for which the NAAQS are set), stratospheric ozone depletion, and risks from air deposition of pesticides and runoff of pesticides into surface waters. Some problems ranked high as health risks and low on ecological and welfare grounds. Among these were hazardous air pollutants, radon and other indoor air pollutants, exposure to consumer products, and exposure to chemicals in the workplace. Conversely, some problems ranked high as ecological and welfare risks but low as health risks: global warming, point and nonpoint sources of water pollution, mining wastes, and physical alteration of aquatic habitats. Others, like groundwater quality and hazardous wastes, ranked medium or low in all the categories.

The result was a ranking of problems that was different from the rankings suggested by laws and appropriations. In the report, EPA noted that its existing program priorities were more consistent with public opinion than with the conclusions of the work groups. Public opinion data at the time showed less concern with indoor air, global warming, and consumer product exposures, which the work groups ranked high in at least two risk categories, than with chemical waste disposal and emergency chemical releases, which the work groups ranked lower.[51] Hazardous wastes, point source water and air pollution, and emergency chemical releases have drawn more public attention and concern, in part because of focusing events like Love Canal and Bhopal.

The Regional Risk Projects *Unfinished Business* was EPA's first systematic attempt to compare and rank problems based on relative risks. Its perspective was national; it did not account for regional/local variations in problems and their importance. In 1987, EPA began regional comparative risk projects in Boston (Region I), Philadelphia (Region III), and Seattle (Region X).[52] The regional projects resembled the national one in the process and methods that were used. The work groups separately ranked health and ecological problems and drew on available data, supplemented by the professional judgments of the participants. There were differences, however, between the regional and national projects. Each region defined its list of problems for analysis, depending on the local conditions and concerns. Over the next few years, all ten regional offices conducted their own comparative risk projects.

The regional risk rankings reflected regional and national concerns about environmental problems. To some extent, regional rankings reflected specifically regional priorities: radon in Region III (Philadelphia), wetlands in Region IV (Atlanta), and problems with industrial emissions in Region V (Chicago), which includes the Great Lakes states. Yet there was consensus across several regions as to what national priorities should be. Among the sources of human health risks, indoor air (including radon) and indoor uses of pesticides ranked high; emergency chemical releases ranked low. As sources of ecological risk, nonpoint sources of water pollution, physical alteration of ecosystems/habitats, and stratospheric ozone ranked high and hazardous waste medium or low.

The regional projects went further than the national project by proposing management strategies for reducing risks that were ranked as priorities. They based the strategies on an evaluation of EPA's legal authority, available technologies, the feasibility and costs of controls, and public perceptions of the problem.

The State-Level Projects A series of state-level projects presents a third level of risk-based planning. Pilot projects in Colorado, Pennsylvania, and Washington State set the model for several state- and (more recently) city-level comparative risk analyses. Based on *Unfinished Business* and the regional projects, the state projects also incorporated more public participation, with public advisory boards, public meetings or hearings, and media outreach. These projects also adopted as a primary goal the completion of an environmental management plan, based on the risk rankings. The use of participation and consensus building distinguishes these from the national EPA project. By November 1993, twenty-six states and eight cities had undertaken comparative risk projects, many of which are still under way.[53]

The SAB Report on Reducing Risk The national, regional, and many of the state-level projects began under Administrators Ruckelshaus and Thomas. After taking office in 1989, William Reilly not only endorsed but expanded risk-based planning at EPA.

Reilly asked the Science Advisory Board to review the data, methodology, and conclusions of *Unfinished Business*, with the purpose of advising him on whether risk-based planning could be used to set broader, long-term environmental priorities. SAB formed its Relative Risk Reduction Strategies Committee, which was divided into three subcommittees: Ecology and Welfare, Human Health, and Strategic Options. SAB set out four goals for itself: to critically review and update *Unfinished Business*; to merge (as much as possible) the evaluations of health, ecological, and welfare risks into combined rankings; to propose strategies for reducing major risks; and to develop a long-term strategy for using risk to set priorities and reduce environmental risks.

The SAB panel offered several recommendations, many relating to the strategies and instruments that will be discussed in chapter 6. However, much of its advice focused on endorsing and refining the risk rankings that EPA, regions, and states had been proposing in previous projects. Although it defined the problems differently, SAB's conclusions did not differ fundamentally from those of the earlier comparative risk projects. The Ecology and Welfare Subcommittee grouped problems as high, medium, or low risks. The higher sources of risk included habitat destruction and alteration (such as the draining of wetlands or deforestation); species extinction and overall losses in biodiversity; global climate change; and stratospheric ozone depletion. Sources of medium ecological risks were pesticides, air toxics, nutrients in water, and acid rain. Ranking relatively low were risks from oil spills, groundwater pollution, and thermal pollution.

The Human Health Subcommittee identified four problem areas as well-documented sources of health risk. Ambient air pollution from stationary and mobile sources includes many substances (ozone, carbon monoxide, benzene, lead, etc.) and many kinds of health effects. Worker exposure to chemicals in industry and agriculture may cause cancer and other problems; the large number of workers exposed to toxics in the workplace makes this a high source of health risk. Indoor pollution from radon, environmental tobacco smoke (i.e., from sidestream smoke), and consumer products (e.g., pesticides, solvents, or formaldehyde in building materials) can be major sources of health risk. Many pollutants in drinking water, among them, lead, chloroform, and disease-causing microorganisms, directly expose a large population to many risks.

Table 9 lists the sources of human health and ecological risk that ranked high in the comparative risk projects EPA sponsored. Although

TABLE 9 PROBLEMS RANKED HIGH IN EPA'S COMPARATIVE RISK PROJECTS

Risk Category	Unfinished Business *Report*	Reducing Risk *Report*	*Regional Projects*
Human health risks	Indoor air/radon	Ambient air pollutants	Indoor air/radon
	Criteria (ambient) air pollutants	Pollution indoors (including radon)	Pesticides
	Household toxics use	Drinking water pollutants	Toxic air pollutants
	Pesticides	Worker exposure to chemicals	Ozone and carbon monoxide
	Drinking water pollutants		
Ecological risks	Stratospheric ozone depletion	Habitat alteration and destruction	Physical alteration of habitat
	Global warming	Stratospheric ozone depletion	Nonpoint source water pollution
	Physical alteration of aquatic habitat	Global climate change	Stratospheric ozone depletion
	Mining, gas, and oil wastes	Losses in biodiversity	Global warming
			Pesticides

NOTE: Different comparative risk projects define problems in different ways.

there are some differences, many problems appear high in all of the rankings. However, it is important to remember that the rankings reflect an evaluation of the "residual" risks associated with each of these problems, that is, the risks that remain given all the pollution control programs now operating in this country. These rankings do not necessarily suggest that programs for dealing with other problems were inappropriate. The rankings do, however, suggest a basis for setting environmental priorities in the United States.

Why Does Risk-based Planning Matter? Altogether, these risk-based planning projects—from *Unfinished Business* to the SAB report to the state- and city-level projects under way now—constitute a deliberate and open effort by a federal agency to shape its policy agenda. Three administrators thought risk-based planning so important that they not only continued but expanded it under their leadership. Why did this happen?

The answer is that there is far more to do out there than there are resources, knowledge, and political capital with which to do it. As the environmental agenda expanded in the 1970s and 1980s, the capacities of EPA and its state and local counterparts were stretched beyond their limits. In addition, as more information about risk became available to policy makers—in large part due to advances in risk assessment—the gaps between what *was* being done and what *should* be done became more apparent. Superfund and the RCRA consumed an increasing chunk of national resources through the 1980s, yet the evidence of risk (beyond those at specific sites) was not strong. At the same time, the potentially higher risks from indoor air pollutants, consumer products, and climate change got far less of society's resources and attention. Problems like nonpoint source water pollution were ranked medium or high, especially as sources of ecological risk, but were poor cousins when compared to U.S. investments in controlling point sources of pollution.

Risk-based planning presents another approach to setting the environmental policy agenda. It allows policy makers to be more systematic than the events, people, and forces that affected their responses in the cases of hazardous waste, Alar, medical waste, and global warming. This effort by EPA to promote a debate about risk, propose methods and approaches for using information about relative risk to set priorities, and affect the relative priority given to problems on its agenda may not transform the larger political process for deciding which problems get attention from Congress, the media, and the public. But it may influence priorities by serving as one of many factors that can shape the policy agenda.

Charles Jones proposes three patterns of agenda setting: (1) the government takes a passive role and reacts to the play of private interests; (2) government defines a process and encourages private parties to take part in setting the agenda; (3) government plays an active role in defining problems and setting goals. The third is distinctive because institutions "systematically review societal events for their effects and set an agenda of government actions."[54] Risk-based planning resembles the second and to some extent the third of these patterns. It is a way to counteract the "chemical of the week" syndrome, in which Congress, the public, and agencies are pulled toward whatever threats loom large on the horizon at any moment, whatever their long-term significance. The slow decline of a wetland may lack the focusing qualities of a cancer scare or a chemical emergency, but it still warrants policy makers' attention.

Risk-based planning is an effort to overcome incrementalism in policy making. Like risk and economic analysis, it can be used to push back the boundaries of rationality. The greatest value of risk-based planning may be to enable agencies to move beyond purely incremental strategies—ones that are reactive, fragmented, and piecemeal—to more actively shape their policy agendas.

CONNECTIONS AND OBSERVATIONS

Understanding how environmental problems are defined and how they may or may not provoke policy responses is critical to understanding environmental policy making. It is worth taking time now to look at some important connections between this and the other chapters.

Analyses give policy makers a measurable basis for defining and comparing problems. Knowing that radon causes 7,000 to 30,000 lung cancer deaths each year or that more than 50 percent of rural drinking water wells have detectable levels of nitrates gives a factual basis for defining problems. Policy makers may not respond to this kind of evidence, but without it they have only intuition or anecdotes with which to identify and compare problems. Certainly, facts are better than intuition or anecdotes on their own, and analysis should be a part of the process for defining problems and setting the policy agenda. Before launching a major national program to reduce lead risks, for example, policy makers would benefit from knowing what exposure occurs, with what effects. It is important to be able to analyze risks in order to define problems and set agendas sensibly.

Economic analysis offers another basis for setting the agenda. With risk as a common metric and with information on likely benefits, agencies can compare problems and establish an analytical basis for giving attention to some problems over others. At times, the evidence on the benefits of acting in response to a problem is persuasive, as the lead case demonstrates. Analysis of the health risks of exposures from lead additives led EPA to phase down and ban lead in gasoline and later to limit lead exposures from other sources, including drinking water. The evidence of health risks, combined with information on the social benefits of reducing lead exposure, made a strong case for action.

Institutions are also a part of the process for defining and comparing problems. Congress, agencies, and courts determine which problems deserve official recognition and thus which deserve a policy response. Often one set of institutions takes a different view of problems than another. Under Reagan and Bush, for example, OMB and the White House were far more skeptical of the need for action on many issues than Congress was. Acid rain was not seen as a problem that needed a policy response (beyond further research) through the Reagan years. The Bush administration was reluctant to accept that global warming was a problem worth acting on. At the same time, environmental groups may create the perception that some condition or scientific prospect is a problem. Skillful public relations and effective use of scientific evidence allow interest groups to draw attention to problems. Risk-based planning offers another way for institutions like EPA and state agencies to promote a structured debate at many levels about priorities.

The relationships between problem definitions and strategies are strong. The way a problem is defined influences the choice of strategies to combat it. Policy makers have defined problems in visible terms and responded with technical fixes—scrubbers on power plants, catalytic converters in cars, triple liners in landfills—rather than with energy conservation, transportation controls, or other comprehensive strategies. Yet interest in such strategies is increasing, as we will see in the next chapter.

CHAPTER SIX

Strategies

"What do successful policy makers actually *do* once a problem has been identified?"[1] When EPA decided to ban most uses of asbestos in the United States in 1989, it followed a standard pattern for responding to a problem. It issued a national rule to reduce and nearly eliminate asbestos use in this country. The rule gave producers and users of asbestos no discretion in deciding when to stop using it. As the regulator, EPA decided what few uses would continue and banned other uses on a fixed timetable. The ban was prospective; it did not affect the asbestos already in place in buildings, automobile brake linings, and elsewhere. It was a classic example of a command-and-control strategy, as direct regulation has come to be known.

Nobody inside or outside EPA was surprised when its response to the asbestos problem took this form. But there were other ways of responding to the risks from asbestos. Liability suits against asbestos producers or users were common; EPA could have let the financial risks of lawsuits discourage future uses. EPA could have informed the public of the hazards of asbestos and let the market determine when asbestos would be used. Or it could have exacted a fee on each ton of asbestos produced or imported into the country to increase the costs of using it and encourage the greater use of substitutes. These and many other responses may have been suitable. But EPA chose regulation, as nearly everyone expected.

Of the array of tools in the environmental policy maker's box, direct regulation has been the most often used in the United States. In some cases, it has probably been the most effective; in others, it has undoubtedly been the least effective. But direct regulation is only one of many ways to respond to problems. The main question that will be addressed in this chapter was posed at the start: What do policy makers do when they determine that something is a problem? From this, other questions follow: What are the options for responding to problems once they emerge on the policy agenda? What alternatives to regulation exist, and what are the reasons for and against their use? Can we respond more efficiently and effectively to environmental problems than we have in the past?

The "emergence of policy problems provokes a search for solutions."[2] There are almost as many solutions as there are problems. In environmental policy as in any other area—social services, criminal law, public health—policy makers tend to draw on solutions they think will be effective in solving a problem. Experience, political needs, administrative culture, legal or budget constraints, and other factors define standard sets of responses—policy tool kits—from which policy makers can draw.

Despite a distinct preference for direct regulation, policy makers in the United States rely on other tools as well. Here are some brief examples of how policy makers have responded to some problems:

Motor vehicle air pollution: Manufacturers are required to use catalytic converters that reduce harmful emissions, and owners are required to pass periodic inspections verifying that the equipment works properly to keep emissions low.

Radon in homes: Information was published and disseminated alerting people to possible risks and encouraging them to test the radon levels in their homes. If there were high levels, policy makers encouraged action to reduce radon to more acceptable levels, using lists of approved contractors.

Inactive hazardous waste sites: A program was established for cleaning up abandoned waste sites, financed either through a government fund or by forcing private parties who contributed to the problem to cover cleanup costs. A goal of restoring the sites to a safe (or safer) condition in order of priority, based on analysis of their relative risks, was set.

Contaminants in drinking water: Maximum allowable levels were set for chemicals that pose health risks. Water suppliers are required to install and operate technologies that keep contamination below those limits. When water suppliers exceed the maximum levels, monitoring and enforcement action are in place.

Loss of wetlands: Farmers, developers, and others proposing to fill in wetlands are required to obtain permits before acting. Permits are granted only when there appear to be no alternatives, when the wetland is of low value, or if there is a plan to restore or create other wetlands to offset the lost acreage.

We can see that policy makers respond to perceived problems in many ways—by setting standards, by issuing permits, by creating funds to cover the costs of cleaning up waste sites. If we look further, we can see other variations in policy makers' responses to problems. Some responses are devised by legislatures, others by agencies; some are implemented at the federal level, others with the states in charge; one response may be aimed at remedying old problems, another at avoiding future ones; one may be pursued strictly within environmental media (air or water), while others cross these media lines to achieve a more integrated response.

Richard Elmore suggests two levels of policy responses to problems. On one level are strategies: "planful, calculated behavior in concert with others whose interests differ."[3] Strategies are made up of policy instruments: "an authoritative choice of means to accomplish a purpose."[4] Groups of policy instruments come together (by design or in some cases by accident) into more or less clearly defined strategies. Some present a well-conceived and rational policy; others emerge piecemeal, as problems evolve or experience accumulates. Policy instruments—regulation, technical assistance, taxes or fees—are a means; strategies are a way to link them.

Elmore's terms are useful in describing how policy makers respond to problems. I especially like the distinction between the two levels of responses, between specific instruments for achieving an end and the ways those instruments come together for action to form strategies. I will simplify the terms even more, however, by defining a strategy as a plan for attacking a problem and a policy instrument as a means for carrying out that attack. We look first at policy instruments and then at strategies, with a closer look at a cross-media lead strategy.

POLICY INSTRUMENTS

Environmental policy makers draw on an array of policy instruments when they respond to problems. I consider three groups of instruments here: information, direct regulation, and market incentives. Direct regulation has traditionally been used the most, but information and market incentives are suited to a new generation of problems and are starting to be used more widely. Each of these groups of policy instruments deserves a careful look. Table 10 lists them, their main features, and examples of each.

INFORMATION

One group of instruments relies on information to carry out policy. The goal is to provide people with information they need to understand and evaluate risks and to take appropriate action to avoid or reduce those risks. Examples are all around us: labels on pesticide containers; warnings on cigarette packages; instructions on paint strippers or other consumer solvents.[5] The importance of information as an instrument is reflected in the recent growth in a field of research and practice known as "risk communication."[6]

Risk communication is a way to influence behavior so as to reduce environmental risks through information and persuasion. Radon is a perfect example of a problem in which risk communication is the best, perhaps the only, way of achieving the desired policy objective. Radon occurs naturally in the soil. Exposures occur largely in housing; they are not the result of the actions by commercial firms or others outside the home. A mandatory testing and remediation program, in which every household would have to test and be forced to take corrective action when radon levels are high, would be politically and economically infeasible. Government's approach has been to work with states and other agencies to get information on the health risks to home owners and induce them to test their radon levels. If the tests reveal radon near or above a recommended "action level" of 4 picocuries per cubic centimeter, people are encouraged to hire a certified contractor to seal off cracks in the foundation, install venting, or take other action. A specific example of a risk communication instrument is EPA's "Citizen's Guide to Radon," which describes the health risks of radon and advises on when to act and what action to take.[7]

Another information approach is technical assistance. EPA and other agencies provide private firms or state and local governments with information, analytical tools, expert advice, and other help for complying with rules or behaving responsibly. As environmental programs have expanded from large industrial sources to cover such diverse sources as small quantity waste generators, underground storage tanks,

TABLE 10 OVERVIEW OF POLICY INSTRUMENTS

Class of Instrument	*Significant Features*	*Examples of Use*
Information and risk communication	Mostly a voluntary approach Good for consumer or life-style risks Less intrusive than other instruments May include information and technical assistance	Radon, household chemicals, community right-to-know
Direct regulation Ambient standards Emission standards Product registration and bans	Direct and visible response to problems Appearance of equal treatment Set clear policy goals or standards Best used for industrial sources of pollution Are intrusive and often inflexible Most common instrument in U.S. policy May be effective but inefficient	NAAQS, Water Quality Standards Effluent guidelines and other technology-based standards Pesticides registration, TSCA product bans
Economic incentives Pollution fees Marketable permits (trading) Deposit-Refund Elimination of market barriers and subsidies	Provide a continuing incentive to reduce pollutants Offer more flexibility for sources Often complement direct regulation Usually more cost-effective than direct regulation Subject of recent interest among policy makers	Carbon fees (not adopted in the U.S.) and VOC fees Emissions trading, lead credit training, acid rain allowances Deposits on lead-acid batteries Changing flood insurance or agricultural subsidy programs

and neighborhood dry cleaners, the need for technical assistance grows. Small firms often lack the resources they need to comply with complex regulatory programs. Here are examples of the kinds of assistance agencies can offer: advising state radon program managers on how to inform home owners; providing computer models to farmers on application rates for fertilizer; designing model recycling programs for local governments to emulate.

A new departure in using information as a policy instrument is Title III in the 1986 reauthorization of Superfund, known as the Emergency

Planning and Community-Right-to-Know Act. It makes industry prepare emergency plans and disclose information on their use and releases of a long list of hazardous chemicals. The intent of the law is to provide communities with information on potential threats to local health and the environment and enable them to work with industry in reducing chemical risks. More than one study has found that the Title III program "provided a powerful incentive for companies to identify and act upon opportunities for reducing accidental and routine releases of hazardous chemicals."[8] In her study of the implementation of right-to-know programs, Susan Hadden argues that Title III can empower citizens. She sees the right to community participation in decisions about chemical uses through a process of informed consent as essential to an effective right-to-know program.[9]

Responding to public concern about harmful chemicals, many states have passed their own requirements for public disclosure. A trendsetter was California's Proposition 65 (the Safe Drinking Water and Toxic Enforcement Act of 1986), which was passed in a statewide initiative.[10] It provides that "no person in the course of doing business shall knowingly and intentionally expose any individual to a chemical known to the state to cause cancer or reproductive toxicity without first giving clear and reasonable warning."[11] Discharges onto the land or into sources of drinking water are covered by the law. It applies to chemicals (some 300 in 1992) on a list recommended by a scientific advisory panel and issued by the governor. The intent of Proposition 65 and similar laws is to force industry to inform the public about chemical risks and allow them to act as they see fit to respond to them.

DIRECT REGULATION

A second group of instruments is direct (or command-and-control) regulation. As noted before, direct regulation has been the mainstay of the environmental policy maker's repertoire. It is what a stethoscope is to the physician or a violin to the symphony conductor; it is hard to imagine environmental protection without direct regulation at its core.

Direct regulation describes a large grouping of instruments that may take several specific forms. We look at three: ambient standards, emission standards, and use restrictions or bans.[12]

Ambient Standards These are applied most fully in the air program. The 1970 act directed EPA to set the NAAQS for the most common conventional air pollutants. These ambient standards define the maximum levels of pollution that are allowed in the outside air. Ambient air levels are affected not only by the volume of the emissions but also by topography, climate, altitude, the location and mix of sources, and daily or seasonal patterns of emissions. Los Angeles is plagued not only by lots of cars and industrial sources but also by warm weather and topography that keeps high levels of ozone in a basin ringed by mountains. For Denver, high carbon monoxide levels are due not only to auto emissions but also the low oxygen at high altitude.

Ambient air quality standards set goals, based on health effects, that each area of the country must meet, whatever the variations in local conditions. They present special problems of monitoring and compliance, because the quality of the ambient air will vary by place and time of day. Air standards are usually defined in terms of parts per million (ppm) and lay out specific protocols for monitoring and compliance.

The water program also relies on ambient standards to some extent. In this case, water quality *standards* are set by states based on water quality *criteria* set by EPA. These criteria are based on ecological rather than on human health risks; aquatic toxicity tests determine what levels of pollutants are acceptable for both freshwater and saltwater environments. Where stream segments or other water bodies exceed the ambient standards, states must set stricter emission limits (for point sources) or management controls (for nonpoint sources) to bring ambient water quality to acceptable levels. Unlike the air quality program, the states have some flexibility in setting ambient water standards, although once they are set, there is a high burden of proof to overcome before a state can make them less stringent. The volume of emissions is only one factor that affects ambient water quality. Some water bodies can absorb higher levels of pollutants than others without harm, due to their chemistry, rate of water flow, or other factors. Monitoring is also an issue; detailed regulations define "mixing zones" and other aspects of the monitoring and compliance program.

Emission Standards These may complement ambient standards or exist on their own as an instrument. Emission limits are the most direct and visible way to reduce pollution levels. They typically are the first instrument policy makers have turned to for results.

Emission standards establish the maximum amounts of pollution that given categories of pollution sources can emit. Like ambient standards, they nearly always are issued as numerical limits—parts per million of pollutants, a limit per unit of fuel or raw material used, total emissions allowed over some time period. They define a standard for emissions from a specific source rather than (as in the case of an ambient standard) for the surrounding environment.

Many important factors affect when and how agencies will set emission standards. One is the distinction between new and old sources. The fact is that environmental policies nearly always emerge after the problems they were designed to address are well established. We instituted controls on large sources of air and water pollution only after the problems became serious enough to warrant a policy response. This time lag between emergence of a problem and a policy response makes it necessary, as a practical matter, to distinguish between old or existing and new sources.

In setting emission standards, policy makers almost always regulate new sources far more stringently than existing ones. New sources are usually held to a "best available technology" standard; those built or starting operation after a given data must install and operate something close to the state-of-the-art technology. Standards for old sources (those existing when the regulatory intervention occurs) vary more but always are less stringent. It is far more costly to refit old sources than to design new sources appropriately; existing firms can better mobilize to resist strict standards; existing problems appear to be easier to accept than new ones. As Peter Huber puts it, "Old risks are risks which society has already embraced or come to tolerate; new risks are those tied to unrealized opportunities."[13]

Not only are emission standards less stringent for existing sources but they are also more likely to be tied to the air quality goals set in the ambient standards. Regulators draw a hypothetical line in the sand. On one side are sources that are not yet constructed or modified, where strict control technologies can be built into the construction process. On the other side are the sources already built and in operation; expensive retrofitting (installing new control technology in old facilities) is necessary to bring them up to current technology standards. The approach has been to impose strict standards on sources on one side of the line (new) and not to retrofit sources on the other side (existing) unless that is necessary to meet the ambient standards. For new sources, emission standards tend to stand alone as an instrument; for existing sources, they are more likely to be used as a complement to ambient standards and as part of a larger strategy.

Use Restrictions or Bans Many environmental problems are the result of products or materials used for a variety of ends, not emissions into the air or discharges into the water. Examples are asbestos and lead. Both are common and pose well-documented risks. One way to deal with such chemicals is to ban their production and use, as EPA did for asbestos. Another way to deal with them is to limit their production or use, as EPA has done with some uses of lead. A ban may be out of the question, because no economically feasible substitutes exist, or because the substitutes may pose more risk than the substance that would be banned. For example, canceling uses of a pesticide shown to pose human health risks may actually increase risks if the substitutes that take its place are more harmful.

Ambient and emission standards are the direct regulatory instruments used under the water and air laws; product bans and restrictions are the regulatory instruments used under the toxics and pesticides laws. Both the TSCA and the FIFRA are product-oriented laws. The TSCA gives EPA authority to act on chemicals that present unreasonable risk. Authority for restricting the production or use of chemicals lies with Section 6 of the TSCA, which permits the use of a range of instruments—from product labels to bans to fees. Again, the distinctions between old and new sources of risk are important. Existing chemicals may be sold and used until EPA finds that they pose unreasonable risks that are not outweighed by their benefits. The burden of proof is on the government. In contrast, new chemicals must pass through a regulatory screen *before* they enter commerce. The burden of proof is on the manufacturer to make a case for the chemical's safety. EPA can deny permission or allow production and use, subject to restrictions set in a regulation.

Pesticides are one of the most highly regulated products in commerce in this country. Here the direct regulatory instrument is a registration process under which EPA licenses a pesticide, then lays out the conditions of sales, application, labeling, and other aspects of its use. The agency controls a pesticide's entry into the market. By not granting permission for a pesticide to enter the market (i.e., by denying the registration), or by revoking the registration for a product that is already on the market, EPA can ban a product. By stating when and how pesticides may be applied, who may apply them, where they may be applied, and what monitoring of use and exposures is necessary, EPA restricts the product's use. It is direct regulation—applied to products and their uses.

So the traditional instrument for achieving environmental policy goals is command-and-control or direct regulation. Agencies set discharge limits, apply these to sources through permits, then verify that the sources meet their limits. Or they control the entry of new products onto the market and define the conditions for use of those products, including perhaps banning them altogether.

There are several advantages to a direct regulatory approach; among them are simplicity, the uniform treatment of sources, and the satisfaction of knowing that each discharger is doing its share (or the most that can be expected) to reduce pollution to desired levels. These advantages surely account for much of the appeal that direct regulation has had in the past. Regulation also is woven into the statutory framework for environmental protection. The Science Advisory Board noted that "EPA looks to conventional regulatory methods for environmental protection because enabling legislation and public expectations push the Agency strongly in that direction."[14] The public expects that those causing pollution will be held accountable. Direct regulation is a visible and often effective (if not efficient) way of meeting this expectation.

Yet the command-and-control approach has several weaknesses as well, some deriving from its apparent strengths.[15] One weakness lies in the apparent simplicity of the technology-based approach. It would seem straightforward to determine what the best available technology is for a given source category emitting a specified list of pollutants. After all, engineers are trained to make these kinds of judgments, agencies can get industry data on processes and finances, and pollution control firms are eager to demonstrate their latest technology as a way to stimulate markets for their products. But consider the obstacles facing an agency when it sets out to establish such limits. There probably are as many economic and technical factors to take into account as there are sources. For any group of emission standards we want to set, there may be several industry categories; we ideally would do a careful analysis before deciding what limits to impose on each. In setting direct regulatory standards, agencies must collect large amounts of data and make judgments about costs, control techniques, and production processes that sources themselves might be better off making.

A second weakness with direct regulation is the *apparent* uniformity of the process for setting emission limits. Remember first that the standards are not as uniform as they appear. New and old sources are treated differently. Smaller sources may be exempted or required to meet less stringent emission limits, due to their lesser ability to absorb the economic burdens of compliance. Even more important, the costs of installing and maintaining controls can vary substantially among sources. As early as 1974, Allen Kneese and Charles Schultze described the results of case studies that demonstrated that large savings were possible when regulators took the differences in marginal control costs among firms into account when setting standards.[16]

A third weakness is that direct regulation may discourage innovation. Regulators decide what is a feasible state-of-the-art technology and require companies to use it. All sources similarly situated must meet the same standards (barring size exemptions). Compliance is a black-and-white issue. If you meet the standards, no matter the margin, you have complied. If you fail, no matter the margin, you have not complied and could face penalties. There is no reward for doing better than directed; there is even something to be lost. Doing better than required by a state-of-the-art standard tells the regulatory agency that it has set the standard too low. So a firm has no continuing incentive to reduce emissions below the permitted levels, even if its costs are low. Clearly, this does not encourage innovation.

MARKET INCENTIVES

Because of weaknesses in direct regulation, policy makers became more interested in economic incentives as an alternative or complementary instrument. Long a favorite of academic economists, incentive instruments have drawn more attention lately from a wide audience, including legislators and EPA officials.[17] The case for market incentives comes from environmental economics. Regulation arises in response to failures in private markets when they treat air, water, land, and other common property resources as a free good. The result is an imbalance between the costs and prices of goods and services sold in markets and the true costs that their production and consumption impose on society. Because the waste-assimilative capacities of common property resources do not carry a financial price to the companies that use them, private markets encourage their overuse, and "the price system conveys the false message that society places no value on clean air and water."[18]

Through direct regulation, government makes companies "internalize" (have to account for) the social costs of their actions. By making firms install control equipment, alter processes, obtain permits, or restrict uses of certain chemicals, agencies limit the use of common property resources—air, land, and water. Regulation is an antidote to the market failures that produced environmental problems in the first place. But regulation is only one way of placing a value on common property resources. Another is to take on defects in the price system more directly, by putting a charge on the use of resources or on the pollution that imposes costs (in the form of environmental damages) on society. This leads us to market incentives as a policy instrument.[19]

Several instruments qualify as market incentives. Most fall into one of five categories: pollution fees, marketable permits, deposit-refund systems, market barrier reductions, and elimination of government subsidies. Each differs from direct regulation by using economic incentives to affect behavior and offering regulated firms some degree of flexibility in deciding how to respond. When they are working properly, incentive instruments induce companies to respond in different ways, depending on the costs of controls, the age of the facility, the nature of the product, or the degree of economic competitiveness in the regulated industry. The result should be a system that equalizes the marginal costs of control across firms and lowers the overall costs of environmental protection to society. I will discuss each category of market incentives in turn (but combine the discussion of market-barrier reductions and subsidies).

Pollution Fees Fees (also known as taxes or charges) make polluters account directly for the social costs of their activities by making them pay for each unit of pollution they produce. While direct regulation makes firms in similar categories control to the same level of stringency, whatever the cost, fees recognize that some firms can control emissions at lower cost than others. Fees induce sources to control emissions or use raw materials in such a way that the marginal cost of control equals the amount of the fee. Anything less costs them more in fees than they would spend in paying for controls; anything more means they pay more to put controls in place than they would in fees. Sources with low costs typically will stay well within their emission limits and so pay less in fees. Firms with high control costs will fall short of the standards but pay higher fees. The result should be the same overall level of reductions but at less cost to society.

There is a certain theoretical appeal in pollution fees, but they have not been used widely. A survey of six countries in the Organization for Economic Cooperation and Development (OECD) found several cases in which fees were used. But the fees were used mostly to raise revenue, not to change behavior or make polluters pay the social costs of the damages they had caused. In only a few cases were the fees linked to the level of pollution generated or the product that was used.[20] They more often were used to cover the administrative costs of government agencies or part of the costs of disposing of waste. So fees often are used to raise revenue, less often to influence behavior. Examples of fees that are designed primarily to raise revenue are trash collection fees (when the charge is not linked with the amounts of trash), taxes on fertilizers imposed in many states (when the amount of the fee is so small that it will have little effect on the amounts used), and federal taxes on chemical feedstocks and oil production that fund the national Superfund program (where imposition of the tax is unrelated to waste practices or responsibility for waste sites).

Yet we can use fees to achieve policy goals, not just raise revenue. To provide an economic incentive, pollution fees must vary according to the amount of pollution released. And they must be set high enough to induce sources to look for new ways to reduce their pollution. Two ways that policy makers can use fees to shape behavior are by charging a fee on the raw materials that are inputs into production processes and by charging a fee on the releases that are outputs from production processes.

Two examples show how we can charge fees on raw materials or inputs. Take lead as an example. It is prudent as a matter of policy not only to reduce exposures to lead in its many forms but also to reduce the total amount of lead mined and entering the environment. One instrument for achieving this would be to levy a fee on the mining, sale, and import of virgin lead. Sellers of virgin domestic or imported lead would be required to collect a fee from purchasers, maintain records of transactions and fees collected, and submit the fee to the government. The aim is to make the production and use of virgin lead more expensive by forcing producers to internalize the social costs of lead they introduce into the environment. The fee gives firms an incentive to recycle lead that is already in use and to find substitutes. As a bonus, income from the fees could be used to sponsor research on lead substitutes that are less damaging to the environment.

Another example of a raw materials fee is a tax on the carbon content of fossil fuels (coal, oil, and natural gas). The amount of carbon in fuels determines how much CO_2 is emitted as a result of the combustion process. The more CO_2 that is emitted, the greater the effects on global warming. The fee would be levied at the point of entry for imported fuel and at the point of production for domestic fuel, such as shipments from coal mines, crude oil as it arrives at refineries, or natural gas received by pipelines. The tax would increase with the fuel's carbon content and provide an incentive for sources to reduce their fuel use overall and to shift to fuels with a lower carbon content. Seen mainly as an instrument for reducing emissions that add to global warming, a carbon fee could also reduce acid rain and improve urban air quality. The effects on emissions could be large: EPA estimated in 1990 that a $25 fee per ton of carbon would reduce emissions by 8 to 17 percent and raise $38 to $50 billion in revenues annually by the year 2000.[21]

Proposals for a carbon fee were considered in the United States in the early 1990s, but they were not adopted. Finland, the Netherlands, Norway, and Sweden have applied fees to fuels based on their carbon content. Denmark, Germany, and Japan plan to introduce such fees.[22]

Another approach is to assess a fee on emissions—on the byproducts or outputs from production rather than the inputs into production. The fee could consist of a flat rate on each unit of emission, but a more effective approach would be to increase the fee as the marginal costs of reducing emissions rise. Faced with an increase in fees for each unit of releases, a firm would "pursue its own interest by reducing pollution by an amount related to the cost of reduction."[23] Companies would be held accountable for the social costs of their emissions by having to pay the fee. Sources with low marginal control costs would remove a higher percentage of their pollution—and pay less in fees—than would sources with high marginal costs, who would be better off paying more in fees.

One example is charging a fee on the emissions of VOCs in high ozone areas. This example also shows how fees can be implemented within a standard regulatory framework. To be workable administratively, the VOC fee would apply only to major sources that emit over one hundred tons a year, which would account for some 10 percent of the VOCs emitted in nonattainment areas. Sources would have to meet the regulatory standards defined in their existing permits—their "baseline" for compliance. But they also would pay a fee on a proportion of the emissions (say 20 percent) within their baseline. A source would be exempt from a fee if it could show that it had reduced its emissions by at least 20 percent below the baseline. Otherwise, it would pay the fee on each unit of VOCs up to the baseline. The amount of the fee per ton could increase over time to provide a continuing incentive for sources to achieve further reductions.[24] The 1990 CAAA authorizes imposition of a VOC fee in certain ozone nonattainment areas.

Policy makers also have considered fees on discharges from wastewater treatment plants, as much for raising revenue as for protecting the environment.[25] One proposal would apply a fee to 189 toxic and 3 conventional pollutants that are discharged into receiving waters and municipal sewers. The pollutants were divided into five categories to reflect their increasing toxicity to aquatic life and human health. The fees per pound of pollutant discharged would vary, depending on the relative toxicity of the five categories of pollutants. They ranged in this proposal from a fraction of a cent for the least toxic category of pollutant to over $60 per pound for the most toxic category. At these levels, the fee could raise some $2 billion annually and give companies an incentive to reduce their discharges. The revenue from the fee could be adjusted by raising or lowering fee levels. But the most socially efficient approach would be to set fees at a level that reflects the marginal social costs of each unit of pollution.

Marketable Permits Another incentive instrument creates artificial markets. Policy makers create such markets by setting a total level for emissions or discharges into the environment, then allocating permits to allow emissions, discharges, or uses of pollutants up to that level.

What if the government placed a limit on the total amount of sulfur oxides that could be emitted each year? It could take a direct regulation approach and assign emission limits to each source and enforce those limits. Or it could allocate a specified amount of allowable emissions to each source and allow them to buy and sell permits within those limits. What if we wanted to cap lead production? Under direct regulation, we would ban or limit uses, as EPA did for asbestos. Marketable permits allow the government to allocate production or use rights to firms, which they can use, sell, or buy from others, depending on the economic value of lead to their business.

Like other incentive instruments, marketable permits aim to make firms accountable for the social costs of their actions. Take a situation in which industrial sources are each granted permits to emit one hundred tons of an air pollutant annually. Some sources will be able to meet this limit easily, because their marginal costs of control are lower than most other sources that emit the pollutant. Others will find that their marginal costs are relatively high; getting their emissions down to required levels could prove to be expensive. It would make sense to allow firms with low marginal costs to reduce emissions below required levels and be able to sell the remaining emission rights to sources that face high costs.

Marketable permits have been used extensively in the United States. The first major use was for emissions trading in the air program. Emissions trading grew out of the restrictions that the Clean Air Act imposed on growth in nonattainment areas during the 1970s. At the time, companies located in nonattainment areas could construct new or modify existing facilities under just two conditions. The first condition was that they install technology that produced the lowest achievable emission rate (LAER). The second condition was to be able to demonstrate, by the time the facility was operating, that existing sources in the area would reduce their emissions by more than enough to offset the new source's emissions. Initially termed the "offset" policy, it became known as emissions trading as EPA expanded its use and offset markets emerged across the country. It was codified in the 1977 CAA and is still part of the program.[26]

The air emissions trading program today actually consists of four separate programs: bubbles, offsets, netting, and banking. The bubble policy allows trading among different points of release of the same pollutant, as long as they are within the same area. The offset policy allows firms located in nonattainment areas to build new facilities if they obtain emission reduction credits from other sources that more than offset their emissions. The banking policy allows firms to save or "bank" emission credits for future use or sale. The netting policy enables sources to modify their existing facilities in some cases without going through the full new source review process that normally applies to new facilities. To do this, sources have to show that the plantwide emissions from the modified facility will not increase much over current levels.

Another application of marketable permits is in the 1990 CAAA, where Congress directed EPA to create a market for sulfur dioxide (SO_2) emissions. The law sets a cap of about 10 million tons annually for SO_2 emissions across the country, to decline to less than 9 million by the end of the century. It assigns emission "allowances" to utilities based on their historic emission levels, emission rates, fuels used, and other factors. Sources that are able to "overcomply" by installing technology, switching to cleaner fuels, or reducing production at old plants and increasing it at newer, cleaner ones can sell their unused allowances to other utilities (presumably with higher costs) that undercomply. "One emitter profits by selling its emissions reduction credit, while the other takes the most cost-effective route to compliance."[27]

The act contains important safeguards. It is designed to produce the same overall emissions reductions. Allowance trading is possible only within specified geographic zones. Continuous emission monitoring and severe penalties for noncompliance (among them a $2,000-per-ton penalty and criminal sanctions) are designed to ensure the integrity of the trading program. Savings could be large; EPA estimates annual savings in the range of $0.7 to $1.0 billion, depending on such factors as the policies of state utility regulators and the willingness of utilities to participate.[28]

The United States has used marketable permits in other situations as well. In the mid-1980s, to reduce the average lead content of gasoline, EPA allowed refiners and importers of gasoline to trade "lead reduction credits." This gave them more flexibility in meeting the limits that EPA had set on the average content of lead in their products. "Refiners and importers that reduced the average lead content of their gasoline below the EPA limit generated credits that could be sold to refiners or importers that exceed the limit."[29] To meet the cuts in CFC production called for in the Montreal Protocol on Substances that Deplete the Ozone Layer (ratified by the United States in 1988), EPA allowed CFC producers to trade their allowances for a steadily declining level of CFC production. Three states—Wisconsin, North Carolina, and Colorado—have established programs that authorize trading of water pollution credits among sources of water discharges.[30]

Deposit-Refund Systems A third incentive instrument puts a surcharge or deposit on products (such as lead-acid batteries) or residuals (such as empty containers) when their improper disposal causes problems. When the product is returned to the seller for recycling or disposal, the deposit is refunded. Users thus have an incentive to return the product or residual. It is a conceptually simple scheme in which users who do not return a product pay the social cost of improper disposal by forfeiting the deposit.

A common application is in the bottle bills that many states have adopted to encourage the reuse and proper disposal of beverage containers. But there are others. A deposit-refund approach has been proposed for such products as lead-acid batteries, used oil, and motor vehicle tires. Once they are returned, the products and their components can be recycled. This eliminates the need for and the risks of disposal and reduces the need for raw materials to make the products in the first place. With lead-acid batteries, for example, a deposit-refund achieves two objectives: it reduces the need to dispose of lead in landfills or incinerators, and it cuts total loadings of lead into the environment by promoting its reuse, thus lowering the demand for virgin lead. Maine and Rhode Island, among other states, have adopted deposit-refund programs for lead batteries.

Market Barriers and Subsidies Often existing government policies present barriers to the operation of market forces and have undesirable effects. In addition, public subsidies of many kinds may support or encourage actions that cause environmental damage. With marketable permits, the policy objective is to create markets. Often, however, the objective of policy has to be to *restore* markets that previous policies have impeded.

Take population growth and commercial development along a coastline. Intensive coastal growth threatens wetlands and other sensitive ecosystems, pollutes coastal waters, and increases the chance for erosion of beaches and other fragile resources. So it would make sense environmentally as a matter of public policy to discourage intensive growth along certain coastlines, especially in ecologically vulnerable areas, or at least not to encourage it. Yet several federal subsidy programs encourage development in sensitive areas by reducing private parties' risk of losses.

An example of such a subsidy program is the National Flood Insurance Program, which insures private parties from losses due to storms, hurricanes, floods, erosion, or changes in sea level. In 1989, the federal government underwrote 2.1 million such policies, covering $170 billion worth of property.[31] Because premiums paid by property owners have been insufficient to cover the program's costs, it has been subsidized with taxpayer funds. By subsidizing property owners against the risks of losses, the program promotes development in coastal areas. By changing the program, government could discourage development in areas that are prone to erosion or otherwise vulnerable. Changes in the program might include (1) requiring new construction to take place outside of erosion-prone areas, (2) discontinuing insurance coverage in high-hazard coastal areas, or (3) increasing the premiums on properties with repeated claims, to encourage people to relocate to new sites instead of rebuilding on existing, vulnerable ones.

Other kinds of government subsidies may contribute to environmental damage. One of the best examples is federal farm crop subsidies, which encourage overproduction, cultivation of marginal lands, and high-input farming methods. A major policy goal recently has been to modify such barriers. The Department of Agriculture's Conservation Reserve Program, for example, requires growers on highly erodible land to develop and implement plans that will minimize soil erosion. Not complying makes them ineligible for federal subsidies, such as commodity payments and loans. The growers' interest in continuing to receive the subsidy gives them an economic incentive to act to achieve the policy goal.[32]

Economic and Environmental Effects of Incentive Instruments Several studies have documented the savings that may come from a greater use of incentive instruments. All the studies done so far show that incentive instruments that allow sources to take a least-cost approach to controlling their pollutants (through market incentives) are more cost effective than direct regulation.

Often the costs of a direct regulatory approach are several times higher than the costs of an economic incentive approach.[33] EPA has estimated that the acid rain allowance trading program will save utilities and consumers between $0.7 and $1.0 billion a year. The total savings to refiners under the lead trading program have been estimated at well over $200 million. The opportunity for lead trading also may have helped refiners reduce the lead content of their gasoline more quickly than they otherwise might have been able to do. Since 1975, the estimated savings to the U.S. economy attributable to air emissions trading range from $5.5 billion to over $12.5 billion. Many advocates of incentive instruments argue that the savings would have been even higher if the trading programs had been designed to be more flexible.

What are the environmental effects of incentive instruments? There has been very little analysis of this issue. The various trading instruments discussed here were designed to produce ratios in emission reductions of greater than one. Each additional unit of pollution allowed under any trade had to be offset by more than one unit of pollution. So emissions trading, if it is properly implemented, should not increase overall emissions.

But there is growing concern about the effects of emissions trading on the *distribution* of pollution. The problem is that the dirtiest companies—those that could be in the market for emissions reduction credits—are often located in poor, minority communities. Although the overall emissions in a region (say, the Los Angeles area, where this issue has come up) would not increase, emissions in certain neighborhoods could go up. "It is fundamentally wrong to promote a regional [emissions] solution that imposes a disproportionate burden on society's poorest members," was how the representative of a Venice, California, environmental advocacy group put it.[34] The efficiency goals of advocates of market incentives may conflict with the equity goals of the emerging environmental justice movement. This conflict will surely receive more attention from policy makers in the coming years.

WHEN IS ONE INSTRUMENT MORE APPROPRIATE THAN ANOTHER?

Instruments have been presented here as policy-neutral means to ends. Like a surgeon's scalpel or the violins in an orchestra, policy instruments can be used well or poorly, as an improvisation or as part of a well-planned set of actions, with more or less definite goals. Drawing on information, direct regulation, or other groups of instruments, we can fashion responses to problems and promote different sets of values. But are there conditions in which one instrument is more useful or effective than others? Are some instruments better suited for certain kinds of problems?

The fit between policy instruments and problems has not been explored very carefully. It is clear that for problems like radon, direct regulation is inappropriate. For highly toxic pollutants or products, where any level of exposure is harmful, direct regulation (either through very strict emission limits or product bans) may be the only feasible strategy. For large industrial sources, some mix of economic incentives and direct regulation may always be necessary. However innovative it is, the acid rain trading program rests squarely on a foundation of direct regulation of emissions. Direct regulation defines the framework within which the trading of acid rain allowances occurs. We should note, in fact, that all of the uses of marketable permits in the United States (emissions trading, lead credit trading, acid rain allowance trading, etc.) took place in conjunction with direct regulation.

On the issue of matching instruments with problems, one study asked, "When should trading schemes be considered—under which circumstances and toward what goals?" It recommends marketable permits in situations when the information needed to set source-specific standards is lacking; there is value in giving polluters incentives to de-

TABLE 11 ISSUES IN DECIDING WHEN TO USE ECONOMIC INCENTIVE INSTRUMENTS

Instrument	*Best Used When*
Pollution fees applied to inputs or raw materials	There is a strong relationship between inputs and environmental damages The goal is to reduce overall loadings into the environment Companies can use substitutes or improve production efficiencies It is feasible to track production, imports, and uses of the material
Pollution fees applied to outputs (emissions, discharges)	Marginal control costs vary among polluters Polluters are able to react to the fee and change behavior It is feasible to monitor emissions There is potential for technological innovation in the industry
Marketable permits	Marginal control costs vary greatly among polluters There is a fixed goal (e.g., total emissions or reductions in emissions) to be achieved Environmental effects are independent of sources and time of emissions (e.g., CFCs) The number of sources is large enough to establish a well-functioning market
Deposit-Refund systems	Serious environmental problems are associated with disposal Recycling and reuse are feasible and profitable Administrative costs are low Retailers, users, and consumers are willing to cooperate

SOURCE: Adapted from *Environmental Policy: How to Apply Economic Instruments* (Paris: Organization for Economic Cooperation and Development, 1991), 99–107.

velop new technologies or control methods; and the marginal costs of complying with uniform standards vary greatly across polluters. Marketable permit instruments may be easiest to apply when the effects of the emissions do not depend much on the location of sources (such as CFC emissions into the atmosphere); the relationship between the time of emission and the environmental effects is not very close; and the number of sources involved "is large enough to establish a well-functioning, competitive market with credits available to trade."[35] Table 11 lists several factors that should be considered in deciding when to use some of the economic incentive instruments that were discussed in this chapter.

Environmental policy could be improved if there were more attention to evaluating the many policy instruments that may be used, their strengths and weaknesses, and the conditions in which they are likely to be most effective. Policy makers could learn a great deal from studies of different policy instruments and their effectiveness in dealing with environmental problems under varied circumstances.

STRATEGIES

Policy instruments offer a menu of options for responding to problems once they earn a place on the agenda. They may be seen as policy-neutral; to select from among information, regulatory, and incentive instruments (or some mix of them) is to decide on a means to an end but not necessarily the end in itself. A decision about the ends of policy and what policy instrument or combination of instruments to use to achieve them brings us to the topic of strategies. This follows on my definition of instruments as a means of attack and strategies as plans for carrying them out.

Strategies for responding to environmental problems vary in their complexity, clarity, and origins. Some are defined in laws and come with detailed schedules and action plans. Others are pieced together by agencies from groups of laws to respond to an emerging problem. At times, strategies specify goals clearly, for example, to reduce lead loadings or allow no net losses in wetlands acreage. At other times, strategies emerge in the form of incremental policy making: in small steps, with evolving or ambiguous goals, while policy makers mutually adjust their plans and actions over time. One strategy may bring a variety of policy instruments together in imaginative ways, with direct regulation here, risk communication there, and marketable permits somewhere else, supplemented with research and education. Others use one instrument—a product ban or emission limit—that constitutes a full response.

By definition, devising a strategy presents policy makers with strategic choices. These choices are not made in a vacuum. They are affected by how the problem is defined, the resources available for dealing with it, and the political and legal constraints that surround the choice. If resources are limited but there is a clear need to respond in some way to a problem, policy makers may use a largely symbolic strategy that gives the appearance of a response. Research, technical assistance, or planning are nonsubstantive and low-cost ways to respond to problems when the financial resources or the political will are lacking. Through the 1980s, the Reagan administration's reluctance to respond to the problem of acid rain led it to push a strategy based on research rather than regulation.

So strategies take many forms and entail several strategic choices. I will consider four such choices here: whether to emphasize pollution control, risk reduction, or pollution prevention; whether to attack a problem nationally or regionally; whether to focus more on avoiding future problems or remedying existing ones; and whether to respond within or across environmental media and policy sectors.

POLLUTION CONTROL, RISK REDUCTION, OR POLLUTION PREVENTION?

One set of strategic choices has assumed a major role in environmental policy making. Imagine three ways of responding to problems. The objective of the first is to prevent pollution. It relies on changes in manufacturing processes, substitutions in raw materials, reuse of by-products—on not generating pollution in the first place. In the second, the objective is to control pollution. Taking existing processes, practices, and materials largely as a given, it applies controls on smokestacks, discharge pipes, or other points of release. In the third, the objective is to reduce risk to humans, wildlife, or resources. The emissions are less important than the effects of exposures to them. Policy is based on the goal of reducing risks that are attributable to pollution.

Environmental policy relies on all three kinds of strategies to some degree. For much of their history, the national programs were based on the second objective, controlling pollution at the point of release. This began to change in the early 1980s, as policy makers turned more to a risk reduction strategy. One reason for this change was the increasing number and variety of sources that were seen to be contributing to environmental problems. As controls imposed on the large point sources in the 1970s and early 1980s began to take effect, the remaining problems were attributed more to the small and diffuse sources. Another reason for greater interest in a risk reduction strategy was that direct regulation was not always the best way to achieve policy goals. Problems like groundwater contamination and radon differed from the earlier problems and required new responses. Finally, the costs and limits of a purely control strategy became more apparent.

By the mid-1980s, risk had become an increasingly attractive analytical perspective for making strategic choices. The return of William Ruckelshaus as EPA's administrator in 1983 brought a change in strategic emphasis.[36] Ruckelshaus sought to develop a strategic conception that would restore credibility to the agency's analyses and decisions, which had been tarnished badly in the Gorsuch years. Developments in risk assessment methods and their greater use by agencies provided the scientific basis for using a more risk-based approach. When the National Academy of Sciences (NAS) issued its 1983 report on managing risk in the federal government, urging more use of risk assessment, it found a willing student in EPA.[37]

Adopting the NAS distinction between risk assessment and risk management, EPA developed guidelines for assessing risks, began to use the concept of relative risk to set priorities, and tried to separate the process of deciding how to *assess* risk from that of deciding how to *manage* it. The first was seen, in the NAS report, as a neutral, technical process that agencies should insulate from political considerations. Decisions about how to manage risk were presented as more value based and political, to be made once the risk assessment was complete. Risk management and the associated goal of risk reduction defined a new strategic orientation.

By the late 1980s, another line of thinking became more prominent. It was apparent that environmental policy in the United States was oriented almost entirely toward controlling pollution at the point of release or managing it at the time of disposal. Indeed, the tendency in this country was to rely heavily on the technical fix, not on altering behavior to reduce what is generated in the first place but applying technical solutions to a smokestack or the end of a pipe. The alternative to pollution control is pollution prevention. Consider the changes that a prevention as opposed to a control emphasis lead us to make: using new kinds of packaging to reduce the paper, plastic, and other materials that are disposed of later; reducing the amount of lead mined and increasing the amounts recycled in secondary smelters to cut overall loadings; adopting farming practices to reduce the use of fertilizers and pesticides; cutting air pollution by reducing the vehicle miles traveled.[38] In each, the objective is to generate less pollution or waste in the first place—with more efficient packaging, less mining and more reuse of a contaminant, more economical use of farm chemicals, and less fuel burned—rather than to take generation of the pollution as a given and apply a technical fix at the points of release.

Today we can see U.S. environmental policy as a combination of these strategic orientations, with occasional tensions among them. Pollution control still is firmly embedded in laws, programs, and practices. Pollution prevention staff at EPA speak of the need for a "culture change" within environmental agencies and industry, in order to move from a control to a prevention orientation.

The risk reduction orientation is alive as well. Risk assessment is, more than ever, a standard part of decision making at EPA. The Science Advisory Board strongly endorsed the use of risk analysis to set both priorities and standards. It makes sense to focus our priorities on the effects of the problems, and risk provides a useful framework for making such comparisons. But prevention makes a great deal of sense as well. Long-term changes in industrial practices, farming methods, and consumer behavior can benefit the environment as well as a company's bottom line.

In coming years, we can expect to see policy makers designing strategies that draw on varied combinations of pollution control, risk reduction, and pollution prevention. Still, they often will need to decide which of the three to stress in specific strategies.

NATIONAL OR REGIONAL IN SCOPE?

Another strategic choice relates to the scope and level of the response to environmental problems. Should the response be uniform and national, with the same goals and standards across the country? Or should it vary regionally and locally?

Several factors affect our choices about the scope for a strategy. One of the most compelling is the relationship in space and time between the causes of problems and their environmental effects. Depletion of the stratospheric ozone layer is a global problem, caused by CFCs and other chemicals that cumulatively destroy ozone molecules in the stratosphere. All emissions of ozone-depleting chemicals contribute to the problem, whatever their geographic origin. The state of Maryland could not solve the problem within its boundaries, for example, even if it totally banned emissions of ozone depletors. The strategy for responding to depletion of the ozone layer must be international in scope.

In contrast, deteriorating water quality in Chesapeake Bay is a regional problem, whose sources lie in the mid-Atlantic area. By cooperating with Virginia, Pennsylvania, and federal agencies, Maryland could devise an effective response to the problem and begin to reverse the decline in water quality in Chesapeake Bay. State and federal officials have in fact developed such a strategy for the bay. A regional strategy is suited to a regional problem.

Environmental policy in this country is a complex mix of local, regional, and national strategies. The pattern in the air, water, and hazardous waste programs has been to set uniform national standards for all new sources and in many cases to set national ambient quality goals for pollutants. The clearest examples are the air quality goals of the CAA or the national, technology-based standards for new sources in the CAA, CWA, and RCRA. For other problems—radon, groundwater, or nonpoint source water pollution—the development of strategies has been left more to state and local policy makers.

For many problems, the choice of the appropriate level or scope for responding is *the* key policy decision. Groundwater policy, for example, is still determined largely at the local and state level. States generally want control over groundwater policy to remain in their hands. Environmentalists and others who favor more stringent, uniform programs push for more national authority. At the national level, views are mixed. EPA officially supports state leadership and limits its role to research, technical assistance, and guidance. However, in specific areas, such as pesticides that may leach into aquifers, EPA has expanded its authority over state groundwater planning activities.

AVOIDANCE OR REMEDIATION?

Another strategic choice is whether to avoid future problems or remedy past ones. Responses to problems necessarily lag behind the evolution of the problems themselves. Policy makers rarely start with a clean slate. Most of the time, environmental policy makers react to problems that already have had consequences.

The need to remedy the effects of past actions is more compelling for some problems and environmental media than for others. Some media, like ambient air or surface water, have an inherent capacity for self-renewal. Problems are self-correcting over a reasonable period of time. Getting emissions of volatile organic chemicals in the Los Angeles basin close to zero now would greatly improve air quality right away. For other problems, the harm or potential for harm already exists and is not easily remedied. Only serious and usually expensive action by policy makers will correct it. For asbestos in buildings, lead in water service pipes, or pesticides in aquifers, the capacities for self-renewal are minimal or nonexistent. A policy response is needed not only to prevent future problems but also to deal with those that already exist.

Asbestos is an excellent illustration. For decades in this country, it was used as a fire-resistant insulator in buildings, ships, pipes, and elsewhere. Although evidence of health risks first appeared some time ago, it was only in the last few decades that public policy began to treat asbestos as a health problem. A first response was to threats posed by asbestos already in place in school buildings, where it often was friable (capable of breaking apart and releasing fibers into the air). The federal government's initial strategy was to make local school officials inspect for friable asbestos and notify the community when there appeared to be a problem. A few years later, Congress went a step further. It directed EPA to develop rules that would require monitoring of indoor air in the schools and corrective action if necessary. EPA then turned from remediation to prevention by issuing the asbestos ban and phase-down rules in 1989. The next round of decisions will focus again on cleaning up the existing problem, as EPA and other agencies grapple with the large quantities of asbestos still in place in commercial and public buildings.[39]

Another example is pesticide leaching into groundwater. Groundwater quality is difficult to monitor. There is anecdotal evidence of problems in parts of the country but no systematic evidence of widespread contamination. Monitoring is expensive, and even thorough monitoring may over- or underestimate rates and levels of contamination. By the time it is clear that there is a problem, it may be too late to do much about it, other than close off the aquifer or take expensive corrective action to treat the water or contain the movement of any contaminants through the aquifer. Alternatives to cleaning up after the fact are implementation of preventive strategies such as requiring users of leaching pesticides to reduce the chances of contamination or canceling the registration of problem pesticides entirely, especially where the groundwater is vulnerable.[40]

WITHIN OR ACROSS ENVIRONMENTAL MEDIA AND POLICY SECTORS?

Recall from earlier discussions how the process for making policy is fragmented by environmental medium and by policy sector. The causes of this fragmentation go back to how problems are defined and organized and even further, to the institutional framework for policy. But just as problems are not confined to specific media, neither should strategies for responding to them be.

And so we reach another strategic choice: Do we respond to problems in the way that institutionally is most feasible, which is to focus within media (air, water) and policy sectors (environment, energy, etc.)? Or do we try to overcome the institutional constraints that separate media and sectors with responses that recognize the cross-media and cross-sectoral nature of problems? The first leads us to artificially fragment problems to make the best use of the existing framework. The second leads to a more integrated approach but brings us face-to-face with the institutional barriers to integrated policy making. The degree of integration by media and sector is an important strategic issue.

Integration can occur at two levels. One is at the sectoral level, through "external" integration (across policy sectors).[41] Such strategies transcend the lines that separate one area of policy from another. Environmental policy is not distinct from but draws on energy, agriculture, transportation, tax, resource management, housing, and trade policy. Tax policy, for example, could be used to promote private investments in less polluting plants and equipment or to influence behavior through fees on emissions. Housing or development policy can be an instrument for controlling growth in ecologically vulnerable areas or removing lead paint and dust from old housing stock. Policies "in sectors not traditionally linked with environmental protection could provide cost-effective environmental benefits that equal or exceed those that can be achieved through more traditional means."[42]

Another level of integration is across media—linking air with water or waste strategies—in "internal" integration (across media but within pollution control programs). This refers to "actions taken across media within environmental policy to prevent release of pollutants and to control residues."[43] At first glance, internal integration might appear easier to achieve than external integration (across policy sectors), because much of the control lies within the reach of a single agency. At the federal and often at the state level, environmental agencies can link strategies across media. Are the chances of setting and carrying out an integrated response to a problem greater when the legal and administrative authority resides under one roof? Sometimes yes and sometimes no. Even in a single agency, there are obstacles to achieving integration. Recall that the institutional setting fosters a fragmented, piecemeal approach. Laws, organization, legislative oversight, constituencies—all combine to fragment rather than to integrate, even within one agency. For a variety of reasons, internal integration has been no easier to achieve than the external kind.

As policy makers have come to view problems in broader, more integrated ways, they have struggled to overcome these obstacles. One means for achieving integration was legislative. When it was passed in 1976, many people saw the TSCA as the framework for linking actions across media on the basis of a chemical or other source of risk. But the act has not lived up to those expectations.

Another approach to policy integration is suggested by the Conservation Foundation's model "Environmental Protection Act," prepared at EPA's request in the late 1980s. It would have replaced the separate laws for air, water, waste, and the other media with a single law. It would have incorporated the existing laws, converted EPA into the Department of the Environment, and set out a common framework for addressing problems across media. The act defined a multimedia standard for "prevention of unreasonable risk" to apply to point and nonpoint sources, mobile sources, and products and chemicals. In applying this standard, agencies would consider six factors: (1) long- and short-term risks to humans and the environment, including cumulative effects of multiple sources; (2) economic costs and their distribution; (3) effects on technological innovation; (4) costs and benefits of any substitutes; (5) the feasibility of the action being implemented; and (6) likely effects on other nations.[44]

Integrating the statutory framework is one approach—from the top down. Another is from the bottom up, at the level of the plant or facility, through integrated facility planning and permitting. An example of an effort to integrate better from the plant level up is a project EPA and the Amoco Corporation conducted at a petroleum refinery in Yorktown, Virginia. Rather than look at environmental releases by medium (air, water, or waste, each on their own), the project involved a comprehensive analysis of releases across all media and the options for reducing them. These options were ranked on the basis of their technical feasibility, their cost, expected reductions in releases that they would achieve, and the risks they presented to people in the plant or the community. The project aimed to identify the barriers to integration at the level of an industrial facility and to assess the feasibility of issuing permits across media, rather than separately for each medium.[45]

Integration at various levels has been something of a holy grail in environmental policy. Achieving it could offer practical gains for policy makers. But, like the holy grail, integration beyond a fairly superficial level—better administrative coordination or organizational design—has proven elusive.

A STRATEGY FOR LEAD

EPA's efforts to develop a cross-media strategy for lead in the early 1990s offers a case study for illustrating many of the points discussed above. In devising a strategy, EPA considered several policy instruments and faced many strategic choices.

Remember that lead is a pervasive, persistent pollutant that causes several health problems. Lead has many uses, and economically feasible substitutes do not currently exist for some of them. Most of the lead to which people are exposed is already in place in the environment: lead paint in old houses, lead residues in soil, lead in drinking water lines, and lead in consumer products. Moreover, we must recognize the political reality that pressures from economic interests who oppose further limits on lead production or use are very strong. A strategy for the problem of lead must tread a fine line if it is to be successful. Strategies are made and carried out within legal, economic, and political constraints that can be influenced but not eliminated.

Some strategies emerge incrementally over time. Other strategies reflect a conscious effort by policy makers to respond to a problem. The lead strategy was the result of a deliberate effort by EPA and other agencies to design a strategy. It responded to the evidence on health risks, especially to children, and to the perception that the government's efforts had to be more focused and better coordinated across agencies and programs. Here is a brief case study of the lead strategy.[46]

STATEMENT OF PROBLEM AND OBJECTIVES

The strategy begins with the health case against lead, discussed in chapter 5. It is a cause of several acute and chronic health problems, from anemia and central nervous system dysfunctions to death. It poses special risks to young children and fetuses by delaying their neurological and physical development. Lead has been linked to hypertension and cardiovascular disease in adults and may have reproductive effects.

As an indicator of levels of risk and the trends in population exposures to lead, the strategy relies on a measure of the level of lead in the blood, that of micrograms (μg) of lead per deciliter (dl) of blood. As recently as 1978, the U.S. Centers for Disease Control (CDC) defined as its "level of concern" blood lead levels of more than 40 micrograms per deciliter (40 µg/dl). As more evidence of the risks of lead exposure has emerged, CDC has lowered this level of concern—to 25 µg/dl in 1991 and to 10 µg/dl since then. Health and environmental officials consider blood lead above this CDC level of concern to be unacceptable, especially for children.

With a numerical measure of health risk, EPA was able to assess trends in blood lead levels in the U.S. population. Mean blood lead levels have gone down by a factor of three or four—from 15 to 20 µg/dl in the late 1970s to about 5 µg/dl in 1991. So mean levels fell from just above the most likely CDC level of concern to well under it. This decline is due largely to the phase-down of lead in gas and to the removal of lead-soldered food cans from domestic production. Similarly, the proportion of children under six with elevated blood levels fell between the late 1970s and 1990: it went from almost 11 percent with levels above 25 µg/dl to 1 percent and from 91 percent with levels above 10 µg/dl to 15 percent. Past efforts to lower exposures are showing results. Yet there are still many children with blood levels above 25 and many more with levels above 10. Because there is no threshold below which lead levels are "safe," even a mean blood lead level of 5 µg/dl is a source of concern.

The goal of the strategy is to reduce lead exposures as much as practicable, particularly to children. It also sets out two specific objectives. The first is to reduce "the incidence of blood lead levels (PbB) above 10 µg/dl (subject to revision in light of the forthcoming CDC report) in children, while taking into account the associated costs and benefits." The aim is to get at lead hot spots where specific kinds of exposure cause elevated blood lead levels and to focus on the most sensitive part of the population, the 15 percent of children above 10 µg/dl. The second objective is to reduce "unacceptable lead exposures that are anticipated to pose risks to children, the general population, or the environment." The first objective was to lower the number of children with elevated (above 10 µg/dl) blood lead levels; the second was to lower the mean blood levels in the population.

ACTION ELEMENTS IN THE STRATEGY

How would the strategy achieve these objectives? Here the concept of policy instruments comes into play. Instruments are a means to an end. Aspects of the problem, resources available for dealing with it, the goals and objectives that policy makers have specified—all determine which instruments will be used and to what ends. EPA's strategy included eight action elements, each of which illustrates the use of one or more groups of policy instruments. I will consider three: implement a lead pollution prevention program; minimize human and environmental exposures by using traditional control mechanisms; and develop and implement a public information and education program. Together, these action elements make up a strategy designed to deal with the political, legal, and other constraints surrounding the lead problem.

> *Implement a pollution prevention program*: This focuses on reducing future exposures from the continued use of lead. It gets at the problem of reducing the total amount of lead released into the environment, rather than the effects of past uses and risks from specific exposures. It includes possible use of two sets of instruments: market incentives to limit or eliminate lead use and exposure; and regulatory instruments like the TSCA to reduce the use of lead in current and future products where the risks outweigh the benefits.
>
> *Minimize exposures through traditional control mechanisms*: This relies on the established use of legal authorities to control lead contamination in all environmental media and to reduce risks from exposures, mostly to lead that is already in place. It depends on direct regulation as the instrument. One aim is to look across media at a lead "cluster" to compare sources of risk (drinking water, air, etc.) and the cost-effectiveness of different regulatory controls.
>
> *Develop and implement public information and education*: The strategy recognizes that educating the public "about sources of lead exposure, how to reduce or avoid exposure, and approaches to preventing additional lead from being introduced into the environment are essential" to its success. It includes a plan for outreach to several audiences through brochures or guidance, seminars and conferences, speeches, activities, and press activities. An example is preparing a training video to help schools monitor lead in drinking water.[47]

STRATEGIC CHOICES

The lead strategy involved many strategic choices. A major strategic choice was whether to stress pollution prevention or risk reduction. A pollution prevention approach goes directly to the problem of total lead loadings. It assumes that any increase of lead in the environment causes damage—now or in the long run. A prevention-based strategy would stress incentives or limits to reduce mining of virgin lead and to encourage the reuse of existing lead, with limits on exposures a secondary objective. A risk-based strategy, in contrast, would look first to exposures and their health and environmental effects. The goal would be to assess and rank sources of risk, then design policy instruments for reducing them when it is cost-effective or cost-beneficial to do so. The concern about reducing overall loadings into the environment would be secondary to that of identifying and reducing sources of risk.

A variety of policy instruments could be used to implement either strategy. But some clear preferences emerge. A prevention approach could draw usefully on incentive instruments. A fee on the mining or import of virgin lead would encourage companies to use substitutes and recycle the lead that is already in use. A deposit on purchases of lead-acid car batteries would induce buyers to return the used batteries to claim a refund, allowing secondary smelting and reuse of the lead. A risk reduction approach would rely more on direct regulation: stricter emission limits on lead smelters, tighter drinking water standards, and restrictions or bans on the use of lead in consumer products.

Lead also involves strategic choices about whether to remedy the effects of past exposures or to avoid future ones. Society will have to pay a price for having used lead in the past—both in the health damages that people suffer and in the costs of the cleanup needed to reduce those damages. The consequences of in-place lead are with us, unless we take steps to remove it from soil or paint where children are exposed. Yet we can act now (and already have acted) to reduce future problems from lead, by reducing the total amount of lead that enters the environment (prevention) or by eliminating exposures that pose risk (risk reduction). Given the costs of doing both and the limits on society's resources for addressing any given problem, what should our strategy emphasize? Although it may at times be politically and economically possible both to remedy past sources of risk and to prevent new ones, there are inevitable constraints on the attention that can be given to a problem. Often policy makers must choose: prevent or remedy?

Lead is very much a cross-sectoral and cross-media problem. To have limited the strategy only to drinking water, or to lead in paint chips, or to air emissions, would have had little effect on the problem. The major single-medium gain in reducing lead risks had already been accomplished, by phasing down lead in gas. The remaining risks could be reduced only by a multimedia strategy. Lead in water supplies is handled by the drinking water program; lead smelters and municipal incinerators are regulated by the air program; lead in landfills and sewage sludge are covered by the waste and water programs; other exposures and uses are subject to TSCA authority and the toxics program. Authority is divided across agencies as well: the Department of Housing and Urban Development (HUD) has authority for lead paint abatement, CDC issues health guidelines, and OSHA focuses on occupational exposures. A cross-media, cross-sectoral strategy was necessary, whatever the difficulties.

STRATEGIES AND POLICY INSTRUMENTS

So what do policy makers do when confronted with a problem? They turn to a fairly standard list of policy instruments and combine them into strategies. Some strategies are legislative, others administrative. Some emerge incrementally, others more systematically. And, of course, some of them work, and others do not. We can pull this discussion of policy instruments and strategies together with some general statements.

First, the way we define problems influences how we design strategies and what instruments we choose. If we define problems narrowly, our responses tend to be narrow as well. Similarly, if we see problems as technical flaws, as American policy often has, we tend to respond with technical fixes. The initial response to automobile pollution was to install catalytic converters—a technical fix—not to limit driving. Even now, the search for the technical fix over more fundamental change continues in the emphasis on alternative fuels. The policy response to ecological risks from pesticides was to cancel the ones that posed the clearest risks, rather than aim to reduce pesticide applications or set aside areas from treatment.

Second, we can break policy responses into two categories: instruments and strategies. Instruments are a means of attack; strategies organize them into a plan of attack. We can see instruments generally as policy-neutral, as a means to an end, although choices among them often imply policy preferences. There are several kinds of instruments. I looked at three that are or could be widely used: information or risk communication, direct regulation, and economic incentives. Environmental policy makers in the United States have relied heavily on direct regulation, although they have used information and incentive instruments more in recent years.

Third, in designing strategies, we make choices. We looked at four strategic choices that environmental policy makers often face: Should they stress pollution prevention, pollution control, or risk reduction? Should they respond nationally or regionally? Should they give priority to remedying past problems or avoiding future ones? Should they design strategies within or among environmental media and policy sectors? In a perfect world, such choices would be unnecessary; we would do what was needed to eliminate a problem. But in the real world of limited resources, multiple and evolving problems, and political constraints, these strategic choices are inevitable. We return to the fact that there are opportunity costs, that we cannot do everything at once, that some problems must be put before others.

The next chapter pulls together the discussion of analyses, institutions, problems, and strategies and returns to the five challenges I posed in the introduction. We then turn to trends in environmental policy that will help in meeting these challenges.

CHAPTER SEVEN

Prospects

Twenty-five years is a long time in environmental policy. Before 1969, there was barely a national program to speak of. What followed was a period of innovation and expansion that changed the way we look at the environment and the government's role in protecting it. From a scattering of goal-setting, grant-making, and planning activities, this country moved to establish a broad and diverse set of regulatory programs. Within a decade, Congress totally revamped and expanded the statutory framework for environmental protection. Today, the tone and substance of policy are very different from what many people might have predicted.

When the national programs were first enacted, the problems appeared to be daunting but solvable. The sources of pollution were large, visible targets for technology-based pollution controls. If only the country could marshal the necessary political will and economic resources, it was thought, the damage the environment had suffered over the last century could be stopped, even reversed. Few better symbols of that belief exist than the zero discharge and fishable/swimmable goals of the 1972 Federal Water Pollution Control Act, or the original Clean Air Act goal of having all parts of the country attain the NAAQS by the mid-1970s.[1]

Time gives us a new perspective; the job is harder and less clear-cut than it first appeared to be. Many problems have stubbornly resisted solution. Ozone levels in Los Angeles are still high. Wetlands continue to be converted for human uses at a rapid rate. Waste disposal becomes more expensive and difficult as landfill capacities fall and incineration remains under political attack. Environmental agencies are strapped for resources in an era of lean government. Environmental and economic goals often come into conflict, although people are beginning to acknowledge that these goals can complement as well as compete with each other. Some problems have been solved, but new ones have emerged. Many old problems have merely taken on new forms. Many problems have not been solved but have been shifted to other environmental media.

Yet it may not be so bad. After all, an elaborate institutional infrastructure is in place. The nation dedicates a good fraction of its annual product to environmental protection. Technology has come to the aid of the environment in many ways. A car sold today is one-twentieth as polluting as its counterpart of 1970. Air pollution controls can remove nearly all contaminants from waste streams, if people are willing to pay the price. New technologies for treating hazardous waste continue to emerge. The federal government has accepted responsibility for cleaning up its weapons facilities, after decades of neglect. An international agreement to eliminate CFC emissions is in force. The Rio summit moved environmental issues to center stage internationally.

On a flight to Los Angeles one day, the woman seated next to me, having learned where I worked, asked how we were doing on the environment. My answer was that there is good news and bad news. On some problems, there has been visible progress.[2] Our more famous polluted waterways, like Lake Erie or the Potomac River, are much improved over what they were just a few decades ago. The percent of the public that is provided with secondary wastewater treatment or better (a high standard of sewage treatment) doubled from 1972 to 1988. Levels of some air pollutants are well below what they were in 1970: sulfur dioxide emissions have fallen by 27 percent, particulates by 63 percent, and lead by 96 percent. At great expense, programs for managing and disposing of hazardous waste have reduced the chances of future Love Canals occurring. Levels of persistent pesticides like DDT in fish and wildlife or of chlordane in humans are well below their 1970 levels.

For other problems, our claims have to be more modest. Take ozone and carbon monoxide (CO). Despite the large increase in vehicle miles traveled in this country since 1970, levels of CO and ozone were stable or declined in most cities. Because of the technology controls that are required for car engines and, more recently, the stricter rules on fuel content, we at least have held our ground on the auto-related air pollutants. Still, about one hundred urban areas do not meet the ozone NAAQS, and some forty fail the CO NAAQS. Levels of another auto-related air pollutant, nitrogen oxides, went up about 7 percent since 1970, but only Los Angeles has been unable to meet the standard in recent years.

For yet another category of problems, it is hard to show much progress at all, and we may have lost ground. Wetlands and coastal waters are an example. Over half of the original wetland acreage in the contiguous United States has been lost. Between 1955 and 1975, over eleven million acres of wetlands were lost; many more are so degraded by pollution and hydrological change that they cannot perform their natural functions. One-third of this country's shellfish beds have been closed due to pollution. One-fourth of the monitored estuaries have elevated levels of toxics. Fish, wildlife, and waterfowl populations have declined in many coastal areas. Globally, CO_2 and other greenhouse emissions continue to increase. If the scientific consensus regarding global warming is accurate, we have only begun to address what could be the greatest environmental failure of all.

So there is good news and bad news on the environment. But the best news may be that it could have been worse: Los Angeles is not Mexico City, the Midwest is not Upper Silesia, and Lake Erie is not the Aral Sea. We at least have avoided the terrible fate that the Lorax could not prevent in the land of the Truffula Trees.

THE ENVIRONMENT, THE ECONOMY, AND DEMOCRACY

More and more, the public is recognizing that economic success and environmental quality are not necessarily incompatible. This is encouraging news for policy makers. The evidence to support this link of environmental with economic progress is more than speculative. A Princeton University study of the relationship between economic development and air pollution found, for example, that pollution "tends to diminish as a country's economy grows beyond a certain level of prosperity."[3] The study compared economic output with pollution levels in dozens of cities around the world. It concluded that pollution was worst in countries with an economic output in the range of $4,000 to $5,000 per person a year and less severe in the much poorer and the richer countries. Pollution tends to increase in the early stages of a nation's development, as a consequence of more industrial activity. Pollution levels then tend to decline as a nation is able to marshal the resources it needs to control pollution and as people achieve the affluence and political awareness to be able to demand environmental quality.

Recent history shows that there can be many patterns in the relationship between economic growth and environmental quality. The United States, Japan, and the nations of Western Europe showed that economic growth is possible—but at a cost to the environment. Yet there still are the resources and the political will in these countries to allow them to act to protect environmental values. Nations of the former Soviet Union and Eastern Europe, however, show a pattern of both environmental devastation *and* economic failure. The central lessons from both experiences may help nations at all stages to achieve more "sustainable" growth.

The concept of "sustainable development" came into use in the late 1980s, as a result of the report of the World Commission on Environment and Development (known as the Brundtland report, after the commission chair). The commission defined it as "development that meets the needs of the present without compromising the ability of future generations to meet their own needs."[4] Although there are many interpretations of just what sustainable development means in practice, we can say generally that it describes policies that factor environmental values into other kinds of policy decisions (agriculture, trade, energy, etc.); account for the long-term value of natural resources, especially nonrenewable resources (like the ozone layer or sensitive ecosystems); and give companies and individuals incentives to protect or promote environmental values. Many of the trends discussed below—the move toward integration, the greater emphasis on preventing pollution, the changes in ways of conducting economic analysis—reflect this growing concern for sustainability.

So it is fair to say that economic development, when it is sustainable, can be good for the environment. And it appears that democracy is good for the environment as well. Consider this fact: Those governments that acted most successfully to restore environmental quality or prevent environmental damage have been the liberal democracies. Whatever faults their critics may find with programs in the United States, the nations of Western Europe, and Japan, the responses to environmental problems in these countries have been far ahead of the responses in most nondemocratic societies. In *The End of History and the Last Man,* Francis Fukuyama observes that "on the whole, democratic political systems reacted much more quickly to the growth of ecological consciousness in the 1960s and 1970s than did the world's dictatorships."[5]

Why this apparent correlation between democracy and concern for environmental quality? Fukuyama explains it in these terms:

> Without a political system that permits local communities to protest the siting of a highly toxic chemical plant in the middle of their communities, without freedom for watchdog organizations to monitor the behavior of companies and enterprises, without a national political leadership sufficiently sensitized that is willing to devote substantial resources to protect the environment, a nation ends up with disasters like Chernobyl, or the dessication of the Aral Sea, or an infant mortality rate in Krakow that is four times the already high Polish national average. . . . Democracies permit participation and therefore feedback, and without feedback, governments will always tend to favor the large enterprise that adds significantly to national wealth, over the long-term interests of dispersed groups of private citizens.[6]

In his widely noted book, Fukuyama argues that the world is evolving toward a collection of societies and political systems that are based on market economics and liberal democracy. If this argument is correct, then the apparent correlation between economic growth (assuming that it is sustainable), democracy, and protection of the environment is good news. It suggests that the economic and political conditions necessary for effective environmental programs will continue to exist, in this country and around the world. And yet, I would argue, our success will depend on our ability to meet the institutional challenges defined at the outset here.

CHALLENGES AND TRENDS IN ENVIRONMENTAL POLICY

This book began with five institutional challenges that environmental policy makers will face in the next few decades:

> *Maintaining democratic values*: The direct effects of environmental decisions on people's lives combine with the strong American traditions of consent and participation to make this an important value in policy making. Yet the difficulty and technical complexity of most environmental issues will require policy makers to develop more effective participatory mechanisms.

Setting the policy agenda: There are far more problems demanding attention than money and political will to solve them. Institutions need to assess the range of problems and focus on the most important ones in the process of setting the environmental policy agenda.

Using social resources effectively: Resources that go to one social goal are unavailable for other social goals. In making policy decisions, policy makers have an obligation to use society's resources prudently. Economic growth and competitiveness must be a part of policy making, or support for environmental programs will suffer and other social needs will go unmet.

Adapting institutions: Our success at dealing with environmental problems will depend on the strength and adaptability of our institutions. The challenges are to integrate programs and policy sectors, enhance the capacities of international institutions, and improve relationships between public and private institutions.

Measuring and evaluating progress: Policy makers need better information on environmental trends and on the results of their efforts at solving environmental problems. Information is often uneven and does not allow us to link measures we have to performance and results.

These challenges overlap in various ways. To set the policy agenda, policy makers need not only good analysis but also mechanisms for consulting with the public. Reliable and acceptable measures of progress support analysis as a way to make more informed, risk-based decisions. Institutions that can integrate across policy sectors will be able to design more comprehensive, cost-effective programs. Better analysis will help agencies determine which policy instruments to use in carrying out what kinds of strategies. Innovative approaches to dealing with industry and the public build stronger democratic institutions and restore public confidence in those institutions. Greater confidence in institutions can lead to better dialogue about analytical methods and their acceptable uses.

These challenges define an institutional agenda for our efforts to improve environmental policy in the years to come. To be sure, the day-to-day work of environmental management will go on. These challenges give us something broader to aim for. So here are some likely trends that also offer solutions—trends we *could* and *should* see in policy in the next few decades.

NOVEL FORMS OF PARTICIPATION

Many people recognize that the adversarial tone and gridlock that can characterize policy making are liabilities. The dispute over the future of nuclear power in this and other countries showed how basic conflicts of values cannot be resolved just by giving people more information or trying to persuade them of the rationality of the experts' analyses.[7] Whether it is in the Washington setting, where representatives of interest groups gather to shape national policy, or locally, for example, when people struggle over the siting of a waste incinerator, we need novel forms of participation to bring citizens, administrators, industry, and others together.

Two mechanisms are especially interesting as we grapple with this need for participation.[8] One is the citizen's panel. Based on the concept of the lay jury, citizen's panels convene groups of nonexperts and nonelites (that is, ordinary citizens) for several days to evaluate data and arguments about an environmental issue. The panel is randomly selected; participants are asked to serve as "value consultants" in a three- to four-day process. A typical panel would hear testimony, question experts, deliberate issues, and agree on recommendations for resolving the issue. Decision makers agree in advance to accept or at least seriously consider the advice of the panel. The premise is that under the right conditions, the lay public can acquire the information and understanding to enable them to evaluate technical policy problems and difficult value choices.

In the United States, the one documented use of a citizen's panel for an environmental issue was for a sewage sludge management project in New Jersey.[9] The panel was convened as part of a research project on participation; the overall goal was to involve the local public in a debate about whether to grant a permit to allow Rutgers University to apply sludge as fertilizer on an experimental farm. The purpose was for the panel to recommend conditions to include in a draft permit, which then would be subject to public hearings. The process by which the panel worked was based on a model first developed in Germany but was modified to conform better to the U.S. political and administrative culture.

The results were mixed. People were pleased to be able to participate early in a decision process, but they wanted a broader array of options than were presented. The panel did not resolve all of the conflicts, but the study group found that the process achieved more "interactive understanding" among industry, citizens, administrators, and experts than is typical in siting decisions.

A second participation mechanism is regulatory negotiation, discussed in chapter 2. Part of a trend to use mediation and negotiation to resolve environmental disputes, regulatory negotiation allows an agency to convene representatives of affected interests in an effort to reach a consensus on a proposed rule. EPA and other agencies have used this process on several rules, but in some cases committees were unable to agree and the negotiations broke down. Even then, the participants said there were benefits to being able to work with other interest groups, discuss the issues, and understand the other participants' points of view.

Citizen's panels bring the lay public into decision making; negotiation uses representatives of recognized interest groups. Citizen's panels are more legitimate for dealing with basic issues suited to their role as value consultants, but negotiation is more suited for implementation issues. A merit of citizen's panels is that they bring the lay public into a decision process and allow citizens to participate directly in decisions. Negotiation, in contrast, is participation for elites, for people whose job it is to represent interests that have a stake in the outcome of the process. Yet both offer a novel approach.

GREATER POLICY INTEGRATION

Having broken environmental policy into small pieces over the years, usually for sound reasons, we now face the challenge of putting it together again. The need to specialize and break complex problems down into more manageable parts explains much of the fragmentation that characterizes the environmental policy process in the United States today. And yet, whether the goal is preventing pollution, protecting a geographic resource (such as Chesapeake Bay), or devising cost-effective strategies for reducing pollutants like lead, policy makers more and more are looking for ways to better integrate policy making.

Integration can be achieved incrementally or through a basic restructuring. An example of an incremental approach was EPA's effort to "cluster" issues on its agenda. The notion of clustering regulations reflected a concern about having to make interrelated decisions in pieces—an air rule here, an RCRA rule there, a water discharge rule somewhere else. The intent was to combine issues that the environmental laws had separated into a set of more coherent clusters that EPA could analyze together. For example, some clusters focused on a resource (groundwater), some on an industrial sector (petroleum refining, oil and gas production), others on a contaminant (lead or nitrogen).[10] The clusters were seen as a way to coordinate related groups of rules; they brought agency staff together to focus on common issues. But the clusters had an even more ambitious purpose: to provide a more integrated framework to compare risks, assess strategies, and make choices. At this they were a mixed success, but they were still an interesting approach to integration.

Clusters illustrate an incremental approach to integration; they take the existing agenda and laws as given. A more radical approach is found in the model integrated statute drafted for EPA by the Conservation Foundation. EPA posed a simple question: If we were to start over by drafting one organic law for all of our environmental programs, how would it look? The model integrated statute that was discussed in chapter 6 is one answer. By using one definition of risk and giving agencies the authority to rank and respond to risks across environmental media, an integrated statute could take us farther along the road to integration.[11]

The search for ways to integrate environmental policy is not just a U.S. phenomenon. The interrelated nature of environmental problems and the many ways in which problems can be defined make integration an issue in any setting. In the United Kingdom, to take an example, pollution control "has traditionally been highly fragmented, involving all levels of government and many different agencies."[12] In an attempt to integrate programs more effectively, the British government created a unified pollution inspectorate within the Department of the Environment in 1987. Its goal was to set standards that represented the "best practicable environmental option" (BPEO) for air emissions, water discharges, and solid waste streams. At a basic level, the BPEO for a given pollution source would be "the optimum combination of available methods of disposal so as to limit damage to the environment to the greatest extent achievable for a reasonable and acceptable total combined cost to industry and to the public purse."[13] Although the concept of the BPEO is appealing in theory, in practice it has been difficult to apply. These same kinds of issues have arisen in the United States and elsewhere in discussions of integration.

Given the political and legal investments in current laws, bureaucratic organization, and program designs, an incremental approach is probably the most feasible way of achieving greater integration in U.S. environmental policy. The many signs of interest in sectoral integration from the Clinton administration, especially in Vice President Gore's grasp of the many links among environmental and other policies, make the odds of achieving greater policy integration—especially *among* federal agencies—higher than they have been in EPA's history.[14]

PREVENTION AS WELL AS CONTROL AND REMEDIATION

Early policy responses to environmental problems took a fairly simple approach: identify the source of the pollution and find a technology to control it. Recently, policy makers have begun to see the need for another strategy, that of preventing pollution before it occurs. The move toward prevention is described at many points in this book. What are some of the key elements in this trend?

EPA's first comprehensive *Pollution Prevention Strategy,* issued in January 1991, gives an idea of what prevention will entail.[15] It outlines actions in four sectors: manufacturing, including the use and production of chemicals; agriculture; municipal waste and wastewater; and transportation and energy. Different approaches for preventing pollution apply in each sector. In the manufacturing sector, some likely approaches would be to change the inputs into a production process (e.g., to substitute nontoxic for toxic feedstocks in making a product) or to change or increase the efficiency of production practices (such as recycling to return waste materials directly to production as raw materials). In the energy sector, a prevention strategy would include measures to increase energy efficiency and rely more on renewable or cleaner energy sources. In the agricultural sector, a strategy would adopt "low-input" practices that use lower levels of pesticides and fertilizers and less water, as well as land management practices that reduce sediment erosion and runoff. What each of these strategies has in common is that the goal is to reduce pollution at the source rather than at the point of discharge.

By the mid-1990s, the concept of pollution prevention had spread throughout U.S. environmental programs. Before 1985, for example, only one state had adopted a law that covered any aspect of pollution prevention. By 1991, over half the states had enacted at least one law on pollution prevention. The contents of these laws vary. Some establish general principles in support of pollution prevention, without much in the way of detail. Others define specific legal obligations, among them, facility plans that companies prepare periodically on efforts to prevent pollution at the source. Like their environmental programs generally, state pollution prevention programs differ. Some programs are "mature, independent, and well-established within the state's environmental hierarchy, and administer a variety of initiatives dealing with pollution prevention." Others are composed of "a coordinator who tries to pull together the pollution prevention aspects of other state environmental programs and whose main job is education about the benefits of pollution prevention."[16]

Private industry, especially the larger companies, has worked to incorporate pollution prevention into its manufacturing and business decisions. Driven by the high costs of waste disposal, a concern over future liabilities, and a growing public concern over the pollution that is generated in many communities, companies have adopted aggressive programs to reduce or eliminate the generation of many wastes. In the late 1980s, for example, Chevron reduced the amount of hazardous waste that it generated by 60 percent and saved more than $10 million in disposal costs. Between 1987 and 1990, Monsanto reduced its hazardous air emissions by 39 percent, through a program of process changes, reuse and recycling of materials, and reengineering of certain industrial processes. A DuPont plant in Texas found that when it adjusted a production process to use less of one raw material, the plant's wastes were cut by two-thirds and it saved $1 million annually. Dow Chemical found that by recycling a toxic solvent used to make an herbicide, it saved some $3 million a year and halved the amount of solvent leaving the plant as waste.[17]

Many companies have implemented their pollution prevention programs as part of an effort to achieve more strategic environmental management. The goal is to integrate environmental management and compliance into a firm's general economic decisions and operations. Government agencies are also trying to take a strategic approach to environmental management, by thinking more in terms of industrial sectors that share certain characteristics than of national rules that apply uniformly to all classes of firms. In its Sustainable Industry Project, for example, EPA's policy office selected three industries (metal finishing, photo-imaging, and a subsector of the plastics industry) as the focus of a sector-based approach. It convened panels of industry experts and others, and with them it identified key factors that affected each industry's environmental management decisions. The panels worked with EPA to prepare action plans. Proposed in the action plans were measures such as changing regulatory requirements that had hampered industry's pollution prevention efforts and promoting development of best management practices by industry. Other suggested measures addressed compliance (because companies in compliance may be at a competitive disadvantage vis-à-vis companies not in compliance) and implementation of technical assistance programs.[18] In addition to creating incentives for pollution prevention, a sector-based approach can promote integration of environmental with other kinds of decisions at the level of the individual company or plant.

MORE USE OF ALTERNATIVES TO DIRECT REGULATION

Slowly but surely, the alternatives and complements to direct regulation have been getting serious attention. What was mostly theory in the 1970s and a series of incremental changes in the 1980s may become an accepted policy instrument in the 1990s and beyond. Emissions trading and the bubble policy were the first uses of incentives. More recently, acid rain trading took the concept further. Are there other policy instruments on the horizon that will make use of economic incentives?

One trend in using economic incentives will be a greater reliance on "social cost pricing," in which the cost of any good or service reflects the full costs to society of providing it. An example is the collection and disposal of municipal waste. In most areas, trash services are covered through general tax revenues or flat fees. Each household pays the same fee, whatever the amount and kind of waste it generates. This is an externality: "a portion of the costs attributable to an individual's action is imposed on the entire community, rather than borne entirely by the individual."[19] Volume-based pricing links service rates to the amount of waste generated; households pay for the direct costs of disposal and (in theory) the social costs of the environmental damage that results from the disposal of the waste.

How would this work in practice? Seattle's approach has received a fair amount of attention. Households subscribe to given levels of service by paying according to the number and size of trash cans that they use each week. In 1990, for example, people paid nearly $11 to use a 19-gallon can and $14 for a 32-gallon can per week. They can buy stickers to cover waste in excess of their standard level of service. Recyclables must be separated and are collected free of charge. Like most cities that began to use volume-based pricing, Seattle initially based rates on the direct costs of collection and disposal, including tipping fees paid at landfills. More recently, however, it has increased fees above the costs of direct service to give households a greater incentive to recycle and reduce the volume of wastes they generate.

With the arrival of the Clinton administration in 1993, there came increased attention to the concept of environmental fees, not only to affect behavior in environmentally beneficial ways but also as sources of revenue. One study argued that setting "green" fees for activities that damage the environment would reduce such damages at lower cost than do current regulatory strategies. Green fees also would allow other taxes to be lowered by 10 percent or more.[20] This study advocates fees for three sets of activities: traffic congestion, solid waste generation, and carbon dioxide emissions. By shifting tax burdens from the "goods" of work and investment (where most taxes now are levied) to the "bads" of congestion, waste, and pollution, the U.S. economy would enjoy a return of $50 to $80 billion a year, in the form of reduced environmental damages and greater economic productivity.

A tax applied to the carbon content of fuels (a carbon tax), for example, could be set at a level that stabilizes emissions at 1990 levels by the year 2000. At this level, some $36 billion in revenue would be generated by the fifth year. Payroll and income tax cuts could be used to offset the relatively higher burdens on low-income households that would occur; other measures would be needed to reduce the adverse economic impacts on coal-producing states like West Virginia and Wyoming. In addition to benefits from lower emissions and lower atmospheric concentrations of CO_2, there could be reductions in sulfur dioxide, nitrogen dioxide, carbon monoxide, particulates, and heavy metals, among others, as well as reduced dependency on oil imports. The goal is to induce users to conserve energy and use the best mix of fuels, because tax rates for different fuels (coal, oil, or natural gas) would vary according to their carbon content. This would be more efficient than relying on traditional, direct regulation.

The study also analyzed pay-by-the-bag systems for pricing municipal waste services and applying variable tolls to drivers on congested roads. Both would offer economic incentives to reduce environmental damages (generating excessive waste or driving on crowded roads), generate revenue, and yield benefits to society.

USE OF ENVIRONMENTAL INDICATORS

The term "indicators" refers to "either direct or indirect measures of environmental quality that can be used to assess the status and trends in the environment's ability to support human and ecological health."[21] The goal is to define technically sound, politically acceptable markers that measure the success of programs and document trends in environmental quality. These are similar to the economic indicators that are used as standard measures of the state of the economy—for inflation, unemployment, growth, and so on.

There are several ways to measure our performance in dealing with environmental problems:

(1) With measures of *activity* by an agency or by sources, such as the number of permits issued, the number of inspections completed, or the kinds and number of control equipment installed by sources of pollution;

(2) With measures of *emissions or discharges* into air or water and onto land, such as the total releases of sulfur oxides in an area in a year or the amount of pesticides applied to crops over a period of several years;

(3) With measures of *ambient concentrations,* such as the levels of ozone in the ambient air or of heavy metals in a river or lake;

(4) With measures of human *exposure* and the exposure of wildlife or ecosystems to environmental contaminants, such as the number of people exposed to levels of air pollution that are above the NAAQS or the number and kinds of fish that are exposed to pollution above the water quality standards in a given area;

(5) With direct measures of *human health or ecological integrity,* such as blood lead levels, the acidity level of lakes and rivers, or species diversity in a forest.

Most environmental agencies have good data on such things as the number of permits they have issued and what enforcement actions they have taken. However useful these measures of agency activity may be in evaluating an agency's performance, though, they tell us little about the effects of these activities on the environment. The fact that a pollution control agency has issued water discharge permits to all industrial facilities in an area does not mean that water quality necessarily will improve. With the second measure—of emissions or discharges—we can begin to keep track of what (if any) changes are occurring in polluters' behavior. Still, reduced air emissions do not necessarily translate into better air quality. The third measure—of ambient concentrations—tells us if air or water quality or levels of contamination in soil are changing but not the effects of those changes on health or ecological resources.

The fourth and fifth measures provide more direct indicators of environmental quality. The fourth allows us to draw conclusions about what harmful contaminants or activities are reaching people, resources, ecosystems, and so on. With the fifth measure, policy makers get direct measures on environmental effects and trends. It is best to think of these five kinds of measures as a continuum, as in figure 5. As we move from the left to the right on the diagram, our ability to document trends in environmental quality improves.

What is the state of environmental indicators? Here are some indicators that were used or were being developed in the early 1990s.

Air: Trends in the ambient concentrations of air pollutants; trends in average global temperatures; amounts of air toxics at points of human exposure; visibility in specified geographic areas.

Water: The percentage of mileage on rivers that supports their "designated" uses; the trophic status of a lake (a measure of ecological health); the number of people drinking water with contaminant levels above the national drinking water standards.

Waste: The number of Superfund sites where cleanup was initiated or progress was made; the volume of materials handled at Superfund sites; the annual increases in the amount of municipal solid waste that is recycled.

Pesticides and toxics: Pesticide residue levels on food; incidents of pesticide poisonings; reuse and recycling of pesticide containers; trends in the industrial emissions of toxic chemicals.[22]

In the early 1990s, EPA began its National Goals Project, which may provide a valuable framework for the further development of environmental indicators. Approximately a dozen broadly defined national environmental goals were established which the country was to achieve early in the next century. Each goal specified an environmental condition; among them were clean air, healthy ecosystems, cleanup and prevention of wastes and toxics, safe food, safe drinking water, and safe indoor air. These goals were proposed as a basis for public debate and participation as well as a set of strategic targets for EPA and other agencies to try to achieve. For each goal, the government was to develop specific objectives and indicators for measuring the nation's progress.[23]

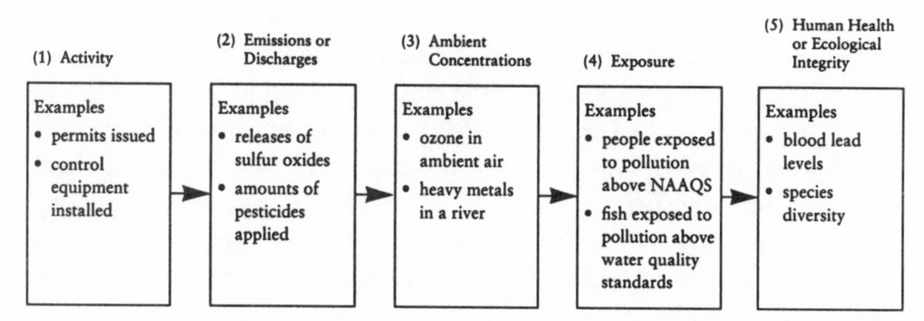

Figure 5. A continuum of measures for environmental indicators. Our ability to document trends in environmental quality improves as we move from left to right.

MORE POLITICALLY LEGITIMATE ANALYSES

A theme in this book is that risk, economic, and other kinds of analysis are appropriate and necessary parts of environmental policy making. I advocate analysis not as a strict decision calculus but as an aid to policy makers in setting priorities and making policy choices.

Yet the role of analysis, especially cost-benefit analysis, is controversial, in large part because of how the Reagan and Bush administrations used it selectively and arbitrarily. Observing OMB and the two administrations' behavior over the last several years, it is clear that cost-benefit analysis was much more a tool in the ideological debate than it was a foundation for fact-based decision making. On the other side of the debate, members of Congress were arguing against the use of almost any kind of economic analysis, in part because the assumptions and methods underlying cost-benefit analysis had become so politically weighted and controversial.

Can we expect to achieve greater political consensus on the use of analysis in environmental policy? Do we reject analysis, continue the current (mostly fruitless) debate, or try to design and use it in ways that are more politically and ethically acceptable to more people? I prefer the third option. Here are two ways to make analysis more acceptable to more people, so that it may be used to support a search for consensus, not obstruct it.

One way is to account better for distributional and equity effects in policy decisions. Traditional welfare economics is oriented toward achieving social efficiency. Efficiency is a worthwhile goal, but at some point efficiency will involve trade-offs with another goal: equity. A cost-benefit analysis could show that a hypothetical policy of concentrating industrial facilities in certain areas would spare 80 percent of the communities from risk but expose the remaining 20 percent to very high levels of risk. The "efficient" policy could be a decision to locate the facilities in the areas where only 20 percent of the people are exposed to risk. This policy could make sense from a cost-benefit point of view, because it provides society in the aggregate with net benefits. But what of the 20 percent bearing a disproportionate share of risks on behalf of society as a whole? And what if the 20 percent consists of minorities?

In simplified form, this is the case behind the environmental justice movement. This movement took off in the late 1980s with a study of the siting of waste facilities across the country, *Dumping in Dixie*. The study concluded that waste facilities were heavily concentrated in minority communities.[24] Based on studies of five such communities (West Dallas, Texas; Institute, West Virginia; Alsen, Louisiana; Emelle, Alabama; and Northwood Manor in Houston, Texas), Robert Bullard found that African-Americans bore a high risk of exposure to contaminants from these facilities and often played no role in the decisions to site waste facilities in their communities. Regardless of their levels of income or social class, minority neighborhoods were more likely to have to tolerate waste facilities than were white ones. Other studies from the 1980s reached similar conclusions.[25] Minorities often cannot escape polluted communities, in part because of "federal housing policies, institutional and individual discrimination in housing markets, federal funding of freeway projects which cut through and disrupt stable minority neighborhoods, and limited incomes."[26]

The growing interest in this issue caused EPA to create its environmental equity work group in the early 1990s. This group led to the agency's first comprehensive effort to assess the risks to minority and low-income communities from environmental pollution. In its report to the EPA administrator, the equity group made several recommendations. The most important was to make equity issues more of a priority on the environmental agenda. Other recommendations flowed from this: to establish and maintain data systems that make it possible to measure risk by race and income; to revise analytical procedures to better characterize risks across populations and communities; to consider distributions of risk in policy decisions; to improve risk communication with minority communities; and to incorporate equity into strategic planning.[27]

In February 1994, President Clinton issued an executive order on environmental justice. The order directs that "each Federal agency shall make achieving environmental justice part of its mission by identifying and addressing, as appropriate, disproportionately high and adverse human health or environmental effects of its programs, policies, and activities on minority populations and low-income populations." The order also directs the EPA administrator to convene an Interagency Working Group on Environmental Justice to coordinate and provide guidance across the government on the specific actions required in the order.[28] In addition, President Clinton's executive order on"Regulatory Planning and Review," issued on September 30, 1993, lists "equity" as a goal in making regulatory decisions.

A second way to make analysis more politically legitimate is to account better for the effects on future generations of policy decisions that are made now. This is particularly an issue when it comes to the long-term effects of policies on ecological resources. As it has been practiced to this point, cost-benefit analysis undervalues such resources as wetlands or species diversity over the long term, and so it allows policy makers to favor consumption of those resources now over their preservation for use or enjoyment in the future. The practice of discounting (i.e., accounting for time preferences in attaching value to resources) is one cause of this tendency to undervalue resources in cost-benefit analyses. Another is the assumption that ecological and other kinds of resources are "fungible" and so can be substituted for each other. Every time we attach a dollar value to (or "monetize") a wetland, a coastal estuary, or the variety of species in a forest, we assume that they are fungible with other resources to which we can attach a dollar value, such as computers or machinery.

Can we be fair to future generations if we attach a time preference to ecological resources?[29] If the value of a wetland in two hundred years is practically zero to this generation—a conclusion that standard cost-benefit analysis would reach using even a very low discount rate—can we ethically justify the loss of that wetland? If a sensitive ecosystem cannot be replaced, how can we justify attaching a dollar value to it at all, given that the only reason we would want to monetize such resources is to be able to compare their value to others that can be substituted for them?

These issues in valuing ecological resources illustrate the general problem of "intergenerational equity" in policy decisions, that is, the problem of the current generation making choices that limit or close off choices for future generations. If we discount future benefits to their present value, as standard cost-benefit analysis would have us do, then we are very likely to shortchange future generations. Similarly, if we assume that irreplaceable resources are fungible with other kinds of resources, when in fact they are not, then we have denied people in the future the right to use or enjoy them. Like environmental justice, this is an issue of equity, only it is equity across generations rather than across social groupings.[30]

Both trends—fairness to minority and low-income groups and fairness to future generations—could reshape environmental policy analysis in the coming years. The environmental justice movement already is well established, as advocates push issues of fairness to minority and low-income groups on the environmental agenda.[31] The stress on ecological values is still more an intellectual trend than a political movement, but there are signs of growing support in the environmental and economics policy communities.[32] The more environmental policy analysts are able to consider the effects of decisions across social groupings and across generations, the more politically legitimate and influential their analyses will be.

VISIONS OF THE FUTURE

Anyone who has tried to do strategic planning has tried to envision what their world will or should be like at some point in the future. Where would you like your company, university, or agency to be ten or twenty years down the road? What vision of the future would you construct given what you want to accomplish?

I end this analysis with two competing visions of the future for environmental policy. One is optimistic, the other pessimistic, though neither is perfectly so. As with any strategic vision, I need to make assumptions. I will make three: that public support for the environment stays at least as strong as it is now; that resources continue to be tight, so that policy must be consistent with economic goals and the need for competitiveness; and that no cataclysmic event comes along—wrenching depression, environmental disaster—to radically change the picture. Given these premises, what visions of the future might we come up with?

AN OPTIMISTIC VISION

An optimistic vision of the future (15–20 years from now) has us moving well past incremental, reactive approaches to problems. Congress and the administration still have their differences, as institutions in a pluralistic system must, but there is broad agreement on the need to set priorities based on shared notions of risk. The Department of Environmental Protection is well established and has parity with State, Defense, and Treasury in the president's Cabinet. The Departments of Energy and Agriculture value environmental quality as highly as they now value maintaining high agricultural or energy production levels. Foreign policy and trade agencies see environmental issues as fundamental to their work. Each house of Congress has created one or two oversight committees to deal with environmental issues across the board—from global warming to air quality to pesticide use.

States have continued to build their credentials as innovators and leaders on the environment, and the serious funding issues of the early 1990s are resolved. The difference between now and then is that states like Mississippi and Utah are more like California, Wisconsin, and New Jersey in their institutional capacities. The federal government has enough confidence in the states to have handed over to them nearly all the responsibility for implementing programs, with minimal oversight from Washington. State and local policies on issues like groundwater protection or full-cost pricing for waste disposal serve as models for the rest of the world.

The Department of Environmental Protection acts under the authority of an organic environmental statute, along with other statutes that Congress might deem necessary to deal with special issues. In addition, statutes governing uses of natural resources, transportation funding and planning, and energy production and use explicitly include environmental protection as a goal. The federal government has greatly improved its ability to look across media and across problems to carry out more integrated analyses of risks, costs, and benefits, as appropriate. Cost-effectiveness analysis is accepted as a useful source of information for policy makers.

Regulated companies are allowed more flexibility in deciding how they achieve pollution goals. They are allowed to propose integrated environmental management plans that enable them to meet their environmental goals more efficiently. The emissions banking and acid rain trading programs have been extended to other air pollutants and to achieving water quality goals. Government strategies rely on market incentives to incorporate the social costs of the production and use of contaminants like lead. Policy makers make greater use of regulatory negotiation and other mechanisms that aim to identify shared values and mutual interests as a basis for reaching consensus. At the state and local levels, experiments with citizen's panels and other novel forms of citizen participation offer lessons in effective democratic policy making.

And, of course, the environment does not look nearly as bad as it might have if these institutional and behavioral changes had not occurred. Only Los Angeles has not met the ozone NAAQS, but it is well on its way. Only a few cities still cannot meet the NAAQS for carbon monoxide. The steady decline in wetlands acreage has been reversed, and state-of-the-art wetlands restoration projects are under way for many high-value wetland areas. The use of pesticides and fertilizers in farming has fallen to a fraction of what it is now, because of the widespread adoption of organic farming methods. Changes in farming practices and better management of land, livestock, and fertilizers have greatly improved water quality in the Great Lakes, Chesapeake Bay, and elsewhere. The volume of municipal waste produced nationally has been greatly reduced, due to more efficient product packaging, recycling, and the existence of efficient markets for recycled materials.

Threats to the global environment have ushered in a new era of international problem solving. Emissions of CFCs and other ozone depletors are down to zero. The richer nations have established a large technical assistance fund to support sustainable growth in the developing countries. A world climate treaty to drastically cut greenhouse gas emissions has been ratified. The treaty commits most countries to a gradual series of emission reductions, and it allows trading among countries to redistribute the cuts if doing so makes economic and political sense. Agreements on regional seas, tropical forestry, and biological diversity under UNEP and other auspices are also yielding concrete environmental gains.

It is, in short, a very promising world for environmental policy. This scenario is not an improbable one. People could debate the merits of the specifics. Yet it is hard to argue that having some or most of this vision fulfilled would not represent a positive trend for the environment.

A MORE PESSIMISTIC VISION

The other vision of the future is not so rosy. The United States has groped its way through almost two more decades of acrimonious debate over priorities. Piecemeal changes in the environmental statutes have further fragmented national policy, and three successive EPA administrators have resigned, citing frustration over micromanagement from too many congressional committees. OMB still is conducting its own guerrilla war to prevent or delay the issuance of rules. As a result, each time Congress reauthorizes a law, it imposes even more deadlines and prescriptive requirements; this further limits agency discretion and frustrates agency efforts to account for the cross-media effects of policies.

An analysis of total national spending on the environment shows that the resources devoted to cleaning up past mistakes far exceed those devoted to preventing future ones. The country falls into a trap, in which the only way to achieve consensus is to focus on problems in which the damages are obvious and the political demands for action overwhelming. Political institutions are so polarized that we find a crazy-quilt pattern: highly stringent rules exist for some problems and almost none for many others, regardless of their relative risk. Despite evidence of the savings that were achieved in the acid rain trading program, policy makers have fallen back on direct regulation exclusively. Risk and economic analyses are used not as a way to evaluate problems, focus resources, or achieve goals more efficiently but as weapons in a continuing debate between industry and government—and within the government itself. State governments try to lead the way but are hampered by limited budgets, lack of direction and assistance from Washington, and a chronic inflexibility in federal rules.

Environmental conditions reflect the institutional failures. Although measures of quality are scarce and a subject of debate, there are some conclusions that people agree on. Ozone still is a nagging air pollution problem; much of southern California rebels at the high costs of controls in the midst of a lagging economy. Although stringent controls exist for all point sources of water pollution, a lack of technical assistance and incentives means that nonpoint sources still are a major cause of water pollution. The losses in wetlands areas continues; little progress is being made in evaluating wetland functions and restoring those of high value.

The international picture is no better. Institutional breakdowns and conflict between advanced and less-developed nations have doomed the long-awaited greenhouse treaty. The United States burns fossil fuel at the same rate as in the past two decades; China aggressively develops its huge coal reserves and cuts corners on emission controls, in part to give it a negotiating lever with the West; Indonesia and Brazil refuse to commit to new policies for preserving their rain forests until the rest of the world offers concessions on the use of fossil fuels. Cracks have even appeared in the stratospheric ozone treaty, because of the limited assistance going from the West to the developing nations.

The irony is that total national spending on the environment has increased more than 50 percent in real terms since 1990; this comes to over 5 percent of a slowly growing GNP. The nation has managed to buy less environmental quality for more money, despite the years of political rhetoric to the contrary.

Which vision of the future is more likely? Take your pick. Both are within reach, depending on the capacity of policy makers and institutions. Neither is out of the question. The challenge is to end up closer to the first vision than to the second.

RATIONALITY AND ENVIRONMENTAL POLICY

Although it is the scientific and technical challenges of environmental policy making that occupy the attention of most people, it is the institutional challenges that may prove to be more formidable over time. Science moves more rapidly than do the collective capacities of institutions or the harried visions of policy makers. We can detect environmental contaminants at the level of parts per billion or better, but we are unable even to begin to unify the many pieces of law that define the goals and framework for our strategies. We can reduce toxic emissions from a waste incinerator by 99 percent and better, but we cannot agree on the role that economics should play in setting standards. Scientists can design complex computer models to predict likely trends in global climate change under many scenarios, while we have been unable to improve much on public hearings as a mechanism for participation.

Making environmental policy is in the end a matter of good politics, effective leadership, creative and adaptable agencies, concerned and involved citizens, good information, and reasoned decision making. Without institutions that are able to connect across problems and set priorities on the basis of some logical premises, policy will be disjointed and resources squandered. Without a sound analytical basis for understanding problems and comparing alternative ways to respond to them, economic and other values will suffer, and the currently high levels of political support for environmental protection could decline. Without an appreciation of the limits of traditional regulation in a world of small and scattered sources, growing international competitiveness, and financially pressed agencies, our ability to deal with problems cannot keep pace with transformations in the problems themselves.

Rationality has been a recurring theme in this discussion. It is easy to make too much of this notion of rationality, to think that it requires full use of quantitative cost and benefit analysis or the reign of technical elites operating in isolation from the distractions of politics or the vagaries of debates over values. Policy making will be more rational not only when social benefits are greater than social costs but when good information about the effects of decisions leads to reasoned debate over the choices that are being made. Rationality may not mean having a fully integrated national environmental policy act that applies the same definition of risk to all problems, but it may mean having mechanisms in place for defining problems more comprehensively and allowing a coherent search for solutions. Being rational means being skeptical about false dichotomies between environmental and economic goals and recognizing that the two may complement each other more than they conflict. Rationality requires that technical experts and administrators accept that the intuitive evaluations of the lay public are as valid as their formal risk and cost-benefit analyses.

Rationality will be an elusive goal in environmental policy. The goal should not be an ideal, textbook form of rationality but a better form of bounded rationality than we have achieved so far. The goal should be institutions that can adapt, integrate, and preserve democratic values; analyses that provide a factual and politically acceptable basis for decision making; a view of problems that recognizes their complexity and interrelationships; an appreciation of risk as a way to set priorities; strategies that can creatively link problems with solutions; and hard choices intelligently made. Achieving this kind of rationality is the best way to fulfill our optimistic vision of the future of environmental policy making.

Notes

CHAPTER ONE: CHALLENGES

1. Dr. Seuss, *The Lorax* (New York: Random House, 1971).
2. On agenda setting, see John W. Kingdon, *Agendas, Alternatives, and Public Policies* (Boston: Little, Brown, 1984).
3. William K. Reilly, "Aiming Before We Shoot: The Quiet Revolution in Environmental Policy," Address to the National Press Club, Washington, D.C., September 26, 1990, 4.
4. Charles O. Jones, *An Introduction to the Study of Public Policy,* 3d ed. (Monterey, Calif.: Brooks/Cole, 1984), 64. The three patterns of agenda setting are described on pp. 62–64.
5. Daniel J. Fiorino, "Environmental Risk and Democratic Process: A Critical Review," *Columbia Journal of Environmental Law* 14 (1989): 501.
6. Seymour Lipset and William Schneider, *The Confidence Gap: Business, Labor, and Government in the Public Mind* (New York: Free Press, 1983). The same authors discuss more recent data in "The Confidence Gap During the Reagan Years, 1981–1987," *Political Science Quarterly* 102 (1987): 1.
7. Discussed further in Daniel J. Fiorino, "Technical and Democratic Values in Risk Analysis," *Risk Analysis* 9 (September 1989): 293–299; and "Citizen Participation and Environmental Risk: A Survey of Institutional Mechanisms," *Science, Technology, and Human Values* 15 (Spring 1990): 226–243.
8. Daniel Mazmanian and David Morell, "The 'NIMBY' Syndrome: Facility Siting and the Failure of Democratic Discourse," in Norman J. Vig and Michael E. Kraft, eds., *Environmental Policy in the 1990s* (Washington, D.C.: CQ Press, 1990), 125–143, on 127.
9. For example, see C. E. Lutrin and A. K. Settle, "The Public and Ecology: The Role of Initiatives in California's Electoral Politics," *Western Political Quarterly* 28 (June 1975): 352–371.
10. The classic work on participation and American political culture is

Gabriel A. Almond and Sidney Verba, *The Civic Culture: Political Attitudes and Democracy in Five Nations* (Princeton: Princeton University Press, 1963). For an excellent discussion of the need for participation in American society, see Benjamin R. Barber, *Strong Democracy: Participatory Politics for a New Age* (Berkeley, Los Angeles, and London: University of California Press, 1984).

11. On the concept of opportunity costs, see Edith Stokey and Richard Zeckhauser, *A Primer for Policy Analysis* (New York: W. W. Norton, 1978), 151–152; E. S. Quade, *Analysis for Public Decisions,* 2d ed. (New York: North-Holland, 1982), 118.

12. The problem of fragmentation in environmental policy has been considered in several works. An early essay is Lynton K. Caldwell, "Environment: A New Focus for Public Policy?" *Public Administration Review* 23 (September 1963): 132–139. A more recent work is Nigel Haigh and Frances Irwin, eds., *Integrated Pollution Control in Europe and North America* (Washington, D.C.: Conservation Foundation, 1990).

13. An example is David Vogel, *National Styles of Regulation: Environmental Policy in Great Britain and the United States* (Ithaca: Cornell University Press, 1986).

14. An early call for such indicators is in J. Clarence Davies III, "The Role of Environmental Indices in Environmental Management," in A. Berry Crawford and Dean F. Peterson, eds., *Environmental Management in the Colorado River Basin* (Logan: Utah State University Press, 1974), 24–36. More recent is U.S. General Accounting Office, *Protecting Human Health and the Environment Through Improved Management* (Washington, D.C., 1988).

15. For a useful introduction to these and other models of the policy process, see Nicholas Henry, *Public Administration and Public Affairs,* 4th ed. (Englewood Cliffs, N.J.: Prentice-Hall, 1989), 292–308.

16. The major early work on the group process model is David B. Truman's *The Governmental Process* (New York: Knopf, 1951). Much of contemporary political science is based on the group model.

17. Marver H. Bernstein, *Regulating Business by Independent Commission* (Princeton: Princeton University Press, 1955).

18. Theodore J. Lowi, *The End of Liberalism: Ideology, Public Policy, and the Crisis of Public Authority* (New York: W. W. Norton, 1969).

19. These contrasts are drawn in Robert B. Reich, "Public Administration and Public Deliberation: An Interpretive Essay," *Yale Law Journal* 94 (1985): 1617–1641.

20. These distinctions are applied to the regulatory process in Daniel J. Fiorino, "Regulatory Negotiation as a Policy Process," *Public Administration Review* 48 (July/August 1988): 764–772.

21. Net-benefit analysis is discussed more in chap. 4. For discussions of its use in environmental regulation, see V. Kerry Smith, *Environmental Policy Under Reagan's Executive Order* (Chapel Hill: University of North Carolina Press, 1984).

22. Graham T. Allison, *Essence of Decision: Explaining the Cuban Missile Crisis* (Boston: Little, Brown, 1971).

23. Herbert Simon, *Administrative Behavior: A Study of Decision-making Processes in Administrative Organizations,* 3d ed. (New York: Free Press, 1976).

24. Charles E. Lindblom, "The Science of Muddling Through," *Public Administration Review* 19 (Spring 1959): 79–88; and *The Policy Making Process* (Englewood Cliffs, N.J.: Prentice-Hall, 1968).

25. Michael D. Cohen, James G. March, and J. P. Olson, "A Garbage Can Model of Organizational Choice," *Administrative Science Quarterly* 17 (March 1972): 1–25.

CHAPTER TWO: INSTITUTIONS I

1. A good discussion of environmental laws is the collection in Paul R. Portney, ed., *Public Policies for Environmental Protection* (Washington, D.C.: Resources for the Future, 1990). An excellent analysis of the passage of the first national laws is J. Clarence Davies III, *The Politics of Pollution* (New York: Pegasus, 1970).

2. A useful source on the NEPA is Richard Liroff, *A National Policy for the Environment: NEPA and Its Aftermath* (Bloomington: Indiana University Press, 1976).

3. For a discussion of the passage of the major laws and the conditions in Congress through the 1970s and 1980s, see Michael E. Kraft, "Environmental Gridlock: Searching for Consensus in Congress," in Vig and Kraft, *Environmental Policy in the 1990s,* 103–124.

4. On efforts to change the Clean Air Act in the early 1980s, see Richard J. Tobin, "Revising the Clean Air Act: Legislative Failure and Administrative Success," in Norman J. Vig and Michael E. Kraft, eds., *Environmental Policy in the 1980s: Reagan's New Agenda* (Washington, D.C.: CQ Press, 1984), 227–249.

5. A good discussion is Brian J. Cook, *Bureaucratic Politics and Regulatory Reform: The EPA and Emissions Trading* (New York: Greenwood Press, 1988).

6. On the water act, see Helen M. Ingram and Dean E. Mann, "Preserving the Clean Water Act: The Appearance of Environmental Victory," in Vig and Kraft, *Environmental Policy in the 1980s,* 251–271.

7. Discussions are Roger C. Dower, "Hazardous Wastes," in Portney, *Public Policies for Environmental Protection,* 151–194; Charles Davis, "Approaches to the Regulation of Hazardous Wastes," *Environmental Law* 18 (1988): 505–535; Steven Cohen, "Defusing the Toxic Time Bomb: Federal Hazardous Waste Programs," in Vig and Kraft, *Environmental Policy in the 1980s,* 273–291.

8. Discussed in Maryann Froehlich, Drusilla J. C. Hufford, and Nancy H. Hammett, "The United States," in Jean-Philippe Barde and David W. Pearce, eds., *Valuing the Environment: Six Case Studies* (London: Earthscan, 1991), 236–265.

9. For an analysis and critique of the Superfund program, see Daniel Mazmanian and David Morell, *Beyond Superfailure: America's Toxics Policy for the 1990s* (Boulder: Westview Press, 1992).

10. On TRI, see Susan G. Hadden, *A Citizen's Right to Know: Risk Communication and Public Policy* (Boulder: Westview Press, 1989).

11. National Academy of Sciences, *Regulating Pesticides in Food: The Delaney Paradox* (Washington, D.C.: National Academy Press, 1987).

12. Arnold W. Reitze, "A Century of Air Pollution Control Law: What's Worked, What's Failed, and What Might Work," *Environmental Law* 21 (1991): 1549–1646, on 1550.

13. Lowi, *The End of Liberalism.*

14. Henry A. Waxman, "An Overview of the Clean Air Act Amendments of 1990," *Environmental Law* 21 (1991): 1721–1816, on 1742.

15. Ibid., 1743. For a lively account of the passage of the act, see Richard E. Cohen, *Washington at Work: Back Rooms and Clean Air* (New York: Macmillan, 1992).

16. See Lynne M. Corn, "The Endangered Species Act: The Storm's Eye," *Bioscience* (October 1990): 637.

17. On the growth of the federal bureaucracy, see James Q. Wilson, "The Rise of the Bureaucratic State," in Francis E. Rourke, ed., *Bureaucratic Power in National Politics,* 3d ed. (Boston: Little, Brown, 1978), 54–78. For an overview of regulation, see Kenneth J. Meier, *Regulation: Politics, Bureaucracy, and Economics* (New York: St. Martin's Press, 1985).

18. On EPA's formation and early years, see Alfred A. Marcus, *Promise and Performance: Choosing and Implementing an Environmental Policy* (Westport, Conn.: Greenwood Press, 1980).

19. Ibid., 101.

20. Examples are Davies, *The Politics of Pollution*; Vig and Kraft, *Environmental Policy in the 1980s*; and Marc K. Landy, Marc J. Roberts, and Stephen R. Thomas, *The Environmental Protection Agency: Asking the Wrong Questions* (New York: Oxford University Press, 1990).

21. Landy, Roberts, and Thomas, *Asking the Wrong Questions,* 38.

22. Analyzed in Martha Derthick and Paul Quirk, *The Politics of Deregulation* (Washington, D.C.: Brookings Institution, 1985). On the general issue of regulatory reform at the time, see Robert E. Litan and William D. Nordhaus, *Reforming Federal Regulation* (New Haven: Yale University Press, 1983).

23. Several discussions of the Reagan years can be found in Vig and Kraft, *Environmental Policy in the 1980s.*

24. Robert Cameron Mitchell, "Public Opinion and Environmental Politics in the 1970s and 1980s," in Vig and Kraft, *Environmental Policy in the 1980s,* 51–74.

25. See Henry C. Krenski and Margaret Corgan Krenski, "Congress Against the President: The Struggle Over the Environment," in Vig and Kraft, *Environmental Policy in the 1980s,* 97–120.

26. Milton Russell, "Environmental Protection for the 1990s and Beyond," *Environment* 29 (September 1987): 12–15, 34–38.

27. *Unfinished Business* and the other comparative risk projects are discussed in more detail in chap. 5.

28. Barbara Rosewicz, "The War Within: Environmental Chief Clashes with New Foe: Deregulation Troops," *Wall Street Journal,* March 27, 1992, A1.

29. For an early assessment, see Timothy Noah, "EPA Chief Carol Browner's First Months in Office Echo the Approach of Her Bush-Era Predecessor," *Wall Street Journal,* July 8, 1993, A14.

30. On the debate over risk and cost-benefit analysis in the Cabinet bill,

see John H. Cushman, Jr., "EPA Critics Get Boost in Congress," *Washington Post,* February 7, 1994, 1.

31. U.S. Environmental Protection Agency, *Summary of the 1995 Budget* (Washington, D.C., February 1994).

32. Marcus, *Promise and Performance,* 107–112.

33. William F. West, "The Growth of Internal Conflict in Administrative Regulation," *Public Administration Review* 48 (July/August 1988): 772–782.

34. For a general analysis of rule making by health and safety regulators, see Gary C. Bryner, *Bureaucratic Discretion: Law and Policy in Federal Regulatory Agencies* (New York: Pergamon Press, 1987). An excellent analysis of science advisers is Sheila Jasanoff, *The Fifth Branch: Science Advisors as Policy Makers* (Cambridge: Harvard University Press, 1990).

35. The *Regulatory Agenda* is published twice a year, in April and October.

36. This discussion is based on the author's own experience in managing the regulatory development process in EPA and on several internal documents, all of which have been available to the public. A document that outlines the process used in the mid-1980s is U.S. Environmental Protection Agency, Office of Standards and Regulations, *A Decision System for Regulatory Development* (Washington, D.C., June 1984). The internal process undergoes constant change, so details of internal reviews and roles vary over time. My intent is to present the goals and features of the process as it worked at least through the early 1990s. My guess is that future approaches will follow the basic model described here, although perhaps with some changes.

37. For a discussion of administrative procedure and the APA, see Paul R. Verkuil, "The Emerging Concept of Administrative Procedure," *Columbia Law Review* 78 (1978): 258–329.

38. On the APA, see Bryner, *Bureaucratic Discretion,* chap. 2.

39. Section 1(b) 46 *Federal Register* 13193 (February 19, 1981).

40. See Cornelius M. Kerwin and Scott R. Furlong, "Time and Rule-making: An Empirical Test of Theory," *Journal of Public Administration Research and Theory* 2 (1992): 113–138. They note a 1990 study citing 600 deadlines for rule making set in environmental laws alone, adding "Congress often enacts these deadlines to placate impatient and powerful interests, with little regard for the feasibility of the imposed timetable" (115).

41. For a detailed and favorable discussion of the options selection process Deputy Administrator Alm adopted, see Thomas O. McGarity, "The Internal Structure of EPA Rulemaking," *Law and Contemporary Problems* 54 (Autumn 1991): 57–111.

42. U.S. Environmental Protection Agency, Office of Policy, Planning, and Evaluation, *EPA's Clusters: A New Approach to Environmental Management* (Washington, D.C., July 1992).

43. On weaknesses in rule making that increased interest in negotiation, see Phillip Harter, "Negotiating Regulations: A Cure for the Malaise," *Georgetown Law Review* 71 (1982): 1–118.

44. This section draws on Fiorino, "Regulatory Negotiation as a Policy Process"; "Citizen Participation and Environmental Risk: A Survey of Institutional Mechanisms"; and "Dimensions of Negotiated Rulemaking: Practical

Constraints and Theoretical Implications," in Miriam K. Mills, ed., *Conflict Resolution and Public Policy* (Westport, Conn.: Greenwood Press, 1990), 141–154.

45. As of January 1994, EPA had completed or begun fourteen regulatory negotiations. The parties reached total or near-total agreement in seven of them and expects agreement in three more that are in progress. Two negotiations ended with agreement on most issues. The remaining two ended well before the parties were able to reach any kind of agreement. I am indebted to Chris Kirtz for providing this information.

CHAPTER THREE: INSTITUTIONS II

1. "Memorandum on Reducing the Burden of Government Regulation," January 28, 1992, published in the *Weekly Compilation of Presidential Documents* 28 (February 17, 1992): 231–268. Reprinted in U.S. Environmental Protection Agency, Office of Policy, Planning, and Evaluation, *Report to the President on the 90-Day Review of Regulations: Overview Volume* (Washington, D.C., May 1992), 56–58. See John E. Yang, "Regulation of Business May Ease," *Washington Post*, January 21, 1991, A1, on what led to the moratorium.

2. R. Shep Melnick, "The Politics of Cost-Benefit Analysis," in P. Brett Hammond and Rob Coppock, eds., *Valuing Health Risks, Costs, and Benefits for Environmental Decision-Making* (Washington, D.C.: National Academy Press, 1990), 23–54, on 32.

3. See Louis Fisher, "Congress as Micromanager of the Executive Branch," in James P. Pffifner, ed., *The Managerial Presidency* (Pacific Grove, Calif.: Brooks/Cole, 1991), 225–237.

4. National Academy of Public Administration, *Congressional Oversight of Regulatory Agencies: The Need to Strike a Balance and Focus on Performance* (Washington, D.C., 1988), 19–24.

5. U.S. Environmental Protection Agency, Office of Congressional and Legislative Affairs, *Status Report* (Washington, D.C., May 31, 1993).

6. National Academy of Public Administration, *Congressional Oversight of Regulatory Agencies*, 12.

7. Richard P. Nathan, *The Plot That Failed: Nixon and the Administrative Presidency* (New York: John Wiley and Sons, 1975).

8. Derthick and Quirk, *Politics of Deregulation*. For an excellent historical analysis, see Marc Allen Eisner, *Regulatory Politics in Transition* (Baltimore: Johns Hopkins University Press, 1993).

9. A very good discussion is Richard D. Waterman, *Presidential Influence and the Administrative State* (Knoxville: University of Tennessee Press, 1989). Another is J. Clarence Davies, "Environmental Institutions and the Reagan Administration," in Vig and Kraft, *Environmental Policy in the 1980s*, 143–160.

10. For an analysis, see Waterman, *Presidential Influence and the Administrative State*, and Robert V. Bartlett, "The Budgetary Process and Environmental Policy," in Vig and Kraft, *Environmental Policy in the 1990s*, 121–142.

11. The classic on incremental budgeting is Aaron Wildavsky's *The New Politics of the Budgetary Process* (Glenview, Ill.: Scott, Foresman, 1988).

12. A useful history of OMB regulatory review is James T. O'Reilly and Phyllis E. Brown, "In Search of Excellence: A Prescription for the Future of OMB Oversight of Rules," *Administrative Law Review* 39 (Fall 1987): 421–444.

13. A good overview is in U.S. Environmental Protection Agency, Office of Policy, Planning, and Evaluation, *EPA's Use of Benefit-Cost Analysis, 1981–1986* (Washington, D.C., August 1987).

14. E.O. 11821, "Inflation Impact Statements," issued November 27, 1974. This order was extended into a requirement for "Economic Impact Statements" (E.O. 11949) on December 31, 1976, which EPA labeled "Economic Impact Analyses."

15. "Improving Government Regulations" (E.O. 12044) issued March 23, 1978.

16. Discussions of E.O. 12291 can be found in Smith, *Environmental Policy Under Reagan's Executive Order*.

17. U.S. Office of Management and Budget, *Regulatory Program of the United States Government, April 1, 1991–March 31, 1992* (Washington, D.C., 1991), from Appendix 4. "Major" rules are those meeting any of several criteria defined in the executive order but usually are those with annual effects on the economy of $100 million or more.

18. U.S. Environmental Protection Agency, Regulatory Development Branch, *OMB Status Report* (Washington, D.C., October 19, 1992).

19. These data were obtained from EPA's Regulatory Development Staff and are available in the author's files.

20. See Dana Priest, "Competitiveness Council Suspected of Unduly Influencing Regulators," *Washington Post,* November 18, 1991, A19; and Philip J. Hilts, "Questions on Role of Quayle Council," *New York Times,* November 19, 1991, B12.

21. Michael Weisskopf, "Wetlands Protection and the Struggle Over Environmental Policy," *Washington Post,* August 9, 1991, A17. Clinton later proposed to modify this wetlands policy to take a more balanced approach. See Stephen Barr, "Clinton to Revise Wetlands Policy," *Washington Post,* August 25, 1993, A1.

22. The special role of the Competitiveness Council revitalized the controversy over the White House's role. An example is the report in Jeffrey H. Birnbaum, "White House Competitiveness Council Provokes Sharp Anger Among Democrats in Congress," *Wall Street Journal,* July 8, 1991, A8.

23. Peter L. Strauss and Cass R. Sunstein, "The Role of the President and OMB in Informal Rulemaking," *Administrative Law Review* 38 (Spring 1986): 181–205, on 186.

24. OIRA staff assigned to EPA rules rarely numbers over half a dozen, although economists from OIRA and OMB budget examiners often take part. An OIRA desk officer usually covers an entire program.

25. On both sides of the OMB debate, see Alan B. Morrison, "OMB Interference with Agency Rulemaking: The Wrong Way to Write a Regulation," *Harvard Law Review* 99 (March 1986): 1059–1074; and Christopher C.

DeMuth and Douglas H. Ginsberg, "White House Review of Agency Rulemaking," *Harvard Law Review* 99 (March 1986): 1075–1088.

26. See the discussion in William A. Nitze, "Improving U.S. Interagency Coordination of International Environmental Policy Development," *Environment* 33 (May 1991): 10–13, 31–37.

27. An excellent discussion of agricultural policies and the environment can be found in the Conservation Foundation's *State of the Environment: A View Toward the Nineties* (Washington, D.C.: Conservation Foundation, 1987), 341–406.

28. National Commission on the Environment, *Choosing a Sustainable Future: The Report of the National Commission on the Environment* (Washington, D.C.: Island Press, 1993), 54.

29. David C. Morrison, "Revamping the Atomic Archipelago," *Government Executive* (June 1991): 46–49. Also see Thomas W. Lippman, "Environmental Law Presses Compliance on U.S. Facilities," *Washington Post,* December 30, 1992, A17. The National Commission on the Environment cites estimates of $70–$110 billion for DOE and $10–$17 billion for DOD, adding "these are probably underestimates" (*Choosing a Sustainable Future,* 54).

30. On American administrative law, see Richard B. Stewart, "The Reformation of American Administrative Law," *Harvard Law Review* 88 (1975): 1667–1813. On the role of the courts, see Harold Levanthal, "Environmental Decisionmaking and Role of the Courts," in H. Floyd Sherrod, ed., *Environmental Law Review—1975* (New York: Clark Boardman, 1975), 545–591.

31. On differences between courts and other institutions, see Donald L. Horowitz, *The Courts and Social Policy* (Washington, D.C.: Brookings Institution, 1977); R. Shep Melnick, *Regulation and the Courts: The Case of the Clean Air Act* (Washington, D.C.: Brookings Institution, 1983); and J. Woodford Howard, Jr., "Adjudication Considered as a Process of Conflict Resolution: A Variation on Separation of Powers," *Journal of Public Law* 18 (1969): 339.

32. This discussion draws on Merrick B. Garland, "Deregulation and Judicial Review," *Harvard Law Review* 98 (1985): 505–591. Also see Verkuil, "The Emerging Concept of Administrative Procedure."

33. Stewart, "The Reformation of American Administrative Law," 1712.

34. Garland, "Deregulation and Judicial Review," 579.

35. Ibid., 586.

36. Rosemary O'Leary, "The Impact of Federal Court Decisions on the Policies and Administration of the U.S. Environmental Protection Agency," *Administrative Law Review* 41 (Fall 1989): 549–574, on 554. Also see Melnick, *Regulation and the Courts.*

37. Illinois v. Gorsuch, 532 F. Supp. 337 (D.D.C. 1981). This case originally was filed as Illinois v. Costle.

38. O'Leary, "Impact of Federal Court Decisions," 561.

39. Ibid.

40. The effluent guideline case was Natural Resources Defense Council v. EPA, 21 Env't. Rep. Cas. (BNA) 1823 (N.D. Cal. 1984).

41. Environmental Defense Fund v. Thomas, 627 F. Supp. 566 (D.D.C. 1986). The case involving DOE facilities is Leaf v. Hodel, 586 F. Supp. 1165 (1984).

42. Corrosion Proof Fittings v. EPA, 89-4596, slip op. (5th Circuit, October 18, 1991). See Michael Weisskopf, "Court Voids EPA Ban on Asbestos," *Washington Post*, October 22, 1991, A19.

43. It gave less deference than the Supreme Court was inclined to give in Chevron U.S.A., Inc. v. Natural Resources Defense Council, 467 U.S. 837 (1984), the leading case in this area.

44. This is discussed in the context of natural gas regulation in Daniel J. Fiorino, "Judicial-Administrative Interaction in Regulatory Policy Making: The Case of the Federal Power Commission," *Administrative Law Review* 28 (1976): 41–88.

45. Melnick, "The Politics of Benefit-Cost Analysis," 39.

46. R. Shep Melnick, "Administrative Law and Bureaucratic Reality," *Administrative Law Review* 44 (Spring 1992): 245–259, on 256.

47. David M. Welborn, "Conjoint Federalism and Environmental Regulation in the United States," *Publius* 18 (Winter 1988): 27–43.

48. Ibid., 31. Welborn places the CAA, FWPCA, FIFRA, Endangered Species Act, SDWA, and hazardous waste parts of the RCRA in the conjoint federalism category.

49. Known, of course, as an FIP (Federal Implementation Plan).

50. On the early patterns under the RCRA, see Ann O'M. Bowman, "Hazardous Waste Management: An Emerging Policy Area within an Emerging Federalism," *Publius* 15 (Winter 1985): 131–144.

51. From James P. Lester, "The New Federalism and Environmental Policy," *Publius* 16 (Winter 1986): 149–165.

52. Ibid., 157. In 1982, thirty-three states depended on federal aid for more than half of the funding for their state environmental programs.

53. A practical guide to state programs is Deborah Hitchcock Jessup, ed., *Guide to State Environmental Programs*, 2d ed. (Washington, D.C.: Bureau of National Affairs, 1990). On four state efforts to innovate in environmental policy, see Barry G. Rabe, *Fragmentation and Integration in State Environmental Management* (Washington, D.C.: Conservation Foundation, 1986).

54. Presented in James P. Lester and Emmet N. Lombard, "The Comparative Analysis of State Environmental Policy," *Natural Resources Journal* 30 (Spring 1990): 301–319, esp. 302–308. The broader set of comparisons was based on an evaluation of state laws, voting records of congressional delegations, and fiscal and other measures. See Christopher J. Duerkson, *Environmental Regulation of State Industrial Siting* (Washington, D.C.: Conservation Foundation, 1983), 224–225.

55. Evelyn Shields, *Funding Environmental Programs: An Examination of Alternatives* (Washington, D.C.: National Governors Association, 1989).

56. "Governors Say Increased Funding Key to Easing Burden of Federal Mandates," *Daily Environment Report*, February 2, 1993, A-5, A-6.

57. On international institutions and the environment, see Lynton Keith Caldwell, *International Environmental Policy: Emergence and Dimensions* (Durham: Duke University Press, 1984); Ved P. Nanda, ed., *World Climate Change: The Role of International Law and Institutions* (Boulder: Westview Press, 1983).

58. William A. Nitze, "The Intergovernmental Panel on Climate Change," *Environment* 31 (January/February 1989): 44.

59. Bruce M. Rich, "The Multilateral Development Banks, Environmental Policy, and the United States," *Ecology Law Quarterly* 12 (1985): 681–745; John Horberry, "The Accountability of Development Assistance Agencies: The Case of Environmental Policy," *Ecology Law Quarterly* 12 (1985): 817–869.

60. World Bank, "Environmental Requirements of the World Bank," *Environmental Professional* 7 (1985): 205–212.

61. Michael Redclift, *Sustainable Development: Exploring the Contradictions* (London: Methuen, 1987). On related issues, see Charles S. Pearson, ed., *Multinational Corporations, Environment, and the Third World* (Durham: Duke University Press, 1987).

62. On the preparations for the summit, see Paul Lewis, "Poor vs. Rich in Rio," *New York Times,* June 3, 1992, A1; or Thomas Kamm, "Some Big Problems Await World Leaders at the Earth Summit," *Wall Street Journal,* May 29, 1992, A1.

63. Both quotes are from Kamm, "Some Big Problems Await World Leaders at the Earth Summit."

64. Rose Gutfeld, "How Bush Achieved Global Warming Pact with Modest Goals," *Wall Street Journal,* May 29, 1992, A1.

65. Cited in Frances Cairncross, *Costing the Earth* (Boston: Harvard Business School Press, 1993), 151.

66. Daniel J. Fiorino, "Institutional Responses to the Emerging Global Environment," in Michael S. Hamilton, ed., *Regulatory Federalism, Natural Resources, and Environmental Management* (Washington, D.C.: American Society for Public Administration, 1990), 189. In April 1993, President Clinton announced that the United States would sign the biodiversity treaty and commit to reducing greenhouse gas emissions to 1990 levels by the year 2000.

67. Discussed in Stewart, "The Reformation of American Administrative Law."

68. Michael Weisskopf, "From Fringe to Political Mainstream," *Washington Post,* April 19, 1990, A1.

69. See, for example, the discussion of Proctor & Gamble's efforts to make its products more environmentally friendly, in Alecia Swasy, "P&G Gets Mixed Marks as It Promotes Green Image but Tries to Shield Brands," *Wall Street Journal,* August 16, 1990, B1.

70. William K. Reilly, "Reflections on the Earth Summit," Memorandum to EPA staff, July 15, 1992, 1.

71. A fuller evaluation of participation mechanisms is in Fiorino, "Citizen Participation and Environmental Risk."

72. See the classic analysis in Almond and Verba, *The Civic Culture.*

73. Francis Rourke, "Bureaucracy in the American Constitutional Order," *Political Science Quarterly* 102 (Spring 1987): 217–232, on 231.

74. Hugh Heclo, "Issue Networks and the Executive Establishment," in Anthony King, ed., *The New American Political System* (Washington, D.C.: American Enterprise Institute, 1978), 87–124, on 102.

75. Ronald Brickman, Sheila Jasanoff, and Thomas Ilgen, *Controlling Chemicals: The Politics of Regulation in Europe and the United States* (Ithaca: Cornell University Press, 1985), 23.

76. Good places to start are Lennart J. Lundqvist, *The Hare and the Tor-*

toise: Clean Air Policies in the United States and Sweden (Ann Arbor: University of Michigan Press, 1980); Vogel, *National Styles of Regulation*; and Sheila Jasanoff, *Risk Management and Political Culture: A Comparative Study of Science in the Policy Context* (Beverly Hills, Calif.: Sage, 1986).

CHAPTER FOUR: ANALYSIS

1. I want to clarify my use of the terms "risk analysis" and "risk assessment" in this discussion. I use risk analysis to describe a general approach to using risk in environmental decision making. I use risk assessment to describe the specific practice of studying and presenting quantitatively the harm that different substances, activities, or technologies may pose to human health or ecological resources. Risk assessment is thus a subcategory of risk analysis.

2. For an intriguing look at risk and risk taking from a psychological and philosophical perspective, see Ralph Keyes, *Chancing It: Why We Take Risks* (Boston: Little, Brown, 1985).

3. There is a large literature on the meaning of risk, estimating risk, and the risks of various kinds of activities. An excellent introduction is Edmund A. C. Crouch and Richard Wilson, *Risk/Benefit Analysis* (Cambridge: Ballinger, 1982). A good collection is Theodore S. Glickman and Michael Gough, eds., *Readings in Risk* (Washington, D.C.: Resources for the Future, 1990).

4. From Richard Wilson, "Analyzing the Daily Risks of Life," *Technology Review* 81 (February 1979): 41–46.

5. James T. Patterson, *The Dread Disease: Cancer and Modern American Culture* (Cambridge: Harvard University Press, 1987), 12.

6. U.S. Environmental Protection Agency, Science Advisory Board, *Reducing Risk: Setting Priorities and Strategies for Environmental Protection* (Washington, D.C., September 1990), 9. *Reducing Risk* comprises four volumes: the volume cited here, which contains the final report of the Relative Risk Reduction Strategies Committee of EPA's Science Advisory Board, and individual volumes for the reports from the committee's Strategic Options, Ecology and Welfare, and Human Health subcommittees.

7. Among the most useful works on risk perception are the following: Mary E. Douglas, *Risk Acceptability According to the Social Sciences* (Beverly Hills, Calif.: Sage, 1985); Baruch Fischhoff, Paul Slovic, and Sarah Lichtenstein, "Lay Fables and Expert Foibles in Judgments About Risk," *American Statistician* 36 (1982): 240–255; Mary Douglas and Aaron Wildavsky, *Risk and Culture: An Essay on the Selection of Technological and Environmental Dangers* (Berkeley, Los Angeles, and London: University of California Press, 1982); Paul Slovic, "Perception of Risk," *Science*, no. 236 (1987): 280–285; and Steve Rayner and Renee Cantor, "How Fair Is Safe Enough? The Cultural Approach to Social Technology Choice," *Risk Analysis* 7 (1987): 3–9.

8. Paul Slovic, Baruch Fischhoff, and Sarah Lichtenstein, "Rating the Risks," in Glickman and Gough, *Readings in Risk*, 61–74.

9. For an excellent discussion of the waste issue, see Susan G. Hadden, "Public Perception of Hazardous Waste," *Risk Analysis* 11 (March 1991): 47–57.

10. J. D. Robinson, M. D. Higgins, and P. K. Bolyard, "Assessing

Environmental Impacts on Health: A Role for Behavioral Science," *Environmental Impact Assessment Review* 4 (1983): 41–53.

11. Fiorino, "Technical and Democratic Values in Risk Analysis."

12. On waste facility siting, a good discussion is chap. 7 in Mazmanian and Morell, *Beyond Superfailure.*

13. For a critique of risk assessment along these lines, see K. S. Shrader-Frechette, *Risk Analysis and Scientific Method: Methodological and Ethical Problems with Evaluating Societal Hazards* (Boston: D. Reidel, 1985).

14. But not to the people who are exposed, for whom population density is not the issue. Small numbers of people may face serious risk that a quantitative analysis may not fully reflect.

15. John J. Cohrssen and Vincent T. Covello, *Risk Analysis: A Guide to Principles and Methods for Analyzing Health and Environmental Risks* (Washington, D.C.: Council on Environmental Quality, 1989), 27.

16. See the summary in U.S. Environmental Protection Agency, *Unfinished Business, Report of the Cancer Risk Work Group* (Washington, D.C., February 1987), B-6 and B-7. For a more detailed discussion, see EPA's "Final Rule for Radon-222 Emissions from Licensed Uranium Mill Tailings: Background Information Document," EPA 520/1-86-009 (August 1986).

17. Discussed in White et al., "A Quantitative Estimate of Leukemia Mortality Associated with Occupational Exposure to Benzene," *Risk Analysis* 2 (September 1982): 195–204.

18. Philip A. Abelson, "Radon Today: The Role of Flimflam in Public Policy," *Regulation* (Fall 1991): 95–100.

19. On both tests and their uses, see Cohrssen and Covello, *Risk Analysis*, 44–48.

20. Recent evidence suggests that at the huge doses given in animal tests, many chemicals destroy cells or chronically irritate tender tissues, causing other cells to divide to replace the lost ones. Dividing cells are more likely to experience the changes in genetic material that lead to cancer. So it may be excessive cell division due to the high doses that are causing the cancers rather than the chemicals. Gina Kolata, "Scientists Question Methods Used in Animal Cancer Tests," *New York Times*, August 31, 1991, A1.

21. On the use of models in multimedia exposure assessments, with municipal waste incinerators as an example, see Jeffrey B. Stevens and Deborah Swackhamer, "Environmental Pollution: A Multimedia Approach to Modeling Human Exposure," *Environmental Science and Technology* 23, no. 10 (1989): 1180–1186.

22. U.S. Environmental Protection Agency, *Environmental Equity: Reducing Risk for All Communities* (Washington, D.C., June 1992). The findings on differential risks are summarized in Vol. 1, *Work Group Report to the Administrator*. Also see Ken Sexton and Yolanda Banks Anderson, "Equity and Environmental Health: Research Issues and Needs," *Toxicology and Industrial Health* 9, special issue (September/October 1993).

23. Susan Okie, "Colon Cancer Risk, Red Meat Linked," *Washington Post*, December 13, 1990, A3.

24. Curtis C. Travis, Samantha A. Richter, Edmund A. C. Crouch, Richard Wilson, and Ernest D. Klema, "Cancer Risk Management: A Review of 132

Federal Regulatory Decisions," *Environmental Science and Technology* 21, no. 5 (1987): 415–420. Most of the decisions were EPA's but also included FDA, OSHA, and CPSC decisions.

25. For a critique of the argument that risk assessment is a "neutral" scientific process that can be separated from "policy" decisions on what agencies should do to reduce risk, see Alice S. Whittemore, "Facts and Values in Risk Analysis for Environmental Toxicants," *Risk Analysis* 3, no. 1 (1983): 23–33.

26. National Academy of Sciences, *Risk Assessment in the Federal Government: Managing the Process* (Washington, D.C.: National Academy Press, 1983). The NAS study also introduced the distinction between the processes of risk assessment and risk management. I do not use that distinction here, because it tends to exaggerate the differences between the scientific and policy aspects of decision making.

27. Joseph Rodricks and Michael R. Taylor, "Application of Risk Assessment to Food Safety Decision Making," in Glickman and Gough, *Readings in Risk*, 143–153, on 150.

28. U.S. Environmental Protection Agency, *Unfinished Business, Report of the Cancer Risk Work Group*, A-2. For more detail on EPA's approach, see its "Guidelines for Carcinogenic Risk Assessment," 51 *Federal Register* 33992 (September 24, 1986).

29. Frances M. Lynn, "The Interplay of Science and Values in Assessing and Regulating Environmental Risks," *Science, Technology, and Human Values* 11 (Spring 1986): 40–50, on 41.

30. Albert L. Nichols and Richard J. Zeckhauser, "The Perils of Prudence: How Conservative Risk Assessments Distort Regulation," *Regulation* (December/November 1986): 13–24. For OMB's criticisms, see Executive Office of the President, *Regulatory Program of the United States Government, April 1, 1990–March 31, 1991* (Washington, D.C., 1990), 13–26.

31. National Academy of Sciences, *Pesticides in the Diets of Infants and Children* (Washington, D.C.: National Academy Press, 1993).

32. On the regulatory budget, see Litan and Nordhaus, *Reforming Federal Regulation*, 133–158.

33. Quade, *Analysis for Public Decisions*, 2d ed., 116.

34. For a discussion of marginal control costs and how they increase with tighter levels of control, see Allen V. Kneese and Charles L. Schulze, *Pollution, Prices, and Public Policy* (Washington, D.C.: Brookings Institution, 1975), 19–22.

35. This discussion draws on U.S. Environmental Protection Agency, Office of Policy Analysis, *Guidelines for Performing Regulatory Impact Analysis*" (Washington, D.C., December 1983).

36. Executive Office of the President, *Regulatory Program of the United States, April 1, 1990–March 31, 1991*, 36.

37. Ibid. OMB gives a succinct discussion of the issue of valuing benefits. Willingness to pay is determined through the "revealed or otherwise discernible preferences of individuals." Markets are a forum for communicating "aggregate valuation for these preferences." Also on p. 36.

38. Discussed in Cairncross, *Costing the Earth*, 44–48. A useful survey of the field is in Maureen L. Cropper and Wallace E. Oates, "Environmental

Economics: A Survey," *Journal of Economic Literature* 30 (June 1992): 675–740. People are willing to pay to improve resources they may not immediately plan to use. There are many types of such "nonuse" values: option value is the willingness to pay to preserve the option of using the resource; existence value is the willingness to pay to know that the quality of a resource is being preserved; and bequest value is knowing that the resource will be preserved for future generations' use.

39. Ann Fisher, Lauraine G. Chestnut, and Daniel M. Violette, "The Value of Reducing Risks of Death: A Note on New Evidence," *Journal of Policy Analysis and Management* 8 (1989): 88–100. The researchers placed more confidence in the estimates at the lower end of this range.

40. Travis et al., "Cancer Risk Management," 419.

41. U.S. Environmental Protection Agency, Office of Policy, Planning, and Evaluation, *Environmental Investments: The Cost of a Clean Environment: A Summary* (Washington, D.C., December 1990). The analysis estimates the direct regulatory implementation and compliance costs of programs. Estimates of the full social costs of the programs would be much higher—and more uncertain. In the report, EPA's projections for the year 2000 give a present and full implementation scenario. The present scenario assumes that present levels of implementation remain the same as in 1987. The full scenario assumes that investments needed to bring about compliance with the ozone standard for air and the fishable/swimmable goals in water for municipal systems are made by the year 2000. The estimates cited here use the full implementation scenario, which gives a slightly higher estimate of total costs for the year 2000 ($160 billion compared to $148 billion) and a slightly higher percentage of GNP (2.8 percent compared to 2.6 percent).

42. U.S. Environmental Protection Agency, *Environmental Investments: The Cost of a Clean Environment* (Washington, D.C., November 1990), 9-1. This is from the full rather than the summary report.

43. Ibid., 4-7. Because these numbers reflect a percentage of GDP rather than GNP, they are slightly different from those cited in table 7. In addition, the U.S. estimates cited here exclude household expenditures, to make them comparable to the European estimates. As a further basis for comparison, we should note that the Netherlands has announced its intention to spend 4 percent of GNP on pollution programs in the future. See William K. Stevens, "2% of GNP Spent by U.S. on Cleanup," *New York Times*, December 23, 1990, 13.

44. A. Myrick Freeman III, "Water Pollution Policy," in Portney, *Public Policies for Environmental Protection*, 97–149.

45. This assumes that one accepts the standard methods for valuing the benefits of environmental programs. Many critics argue that conventional economics undervalues many benefits, especially ecological ones. I consider this further at the end of this chapter and in the concluding chapter. The cost estimates could also be biased upward; see ibid., 126–127.

46. Paul R. Portney, "Economics and the Clean Air Act," *Journal of Economic Perspectives* 4 (Fall 1990): 173–181.

47. Robert W. Hahn and John A. Hird, "The Costs and Benefits of Regu-

lation: A Review and Synthesis," *Yale Journal on Regulation* 8 (Winter 1991): 233–278, table 2. The cost estimates are slightly lower than EPA's, because of different assumptions. The point is the comparison between costs and benefits. To take two other extremes, they estimate the costs of highway safety programs at $6.4–$9.0 billion with benefits of $25–$47 billion; and the costs of OSHA programs as $8.5–$9.0 billion with benefits that are "negligible."

48. An early discussion of economic impact analysis is Robert H. Haveman and V. Kerry Smith, "Investment, Inflation, and Unemployment," in Paul R. Portney, ed., *Current Issues in U.S. Environmental Policy* (Baltimore: Johns Hopkins University Press, 1978), 164–200.

49. For critiques of cost-benefit analysis, see Robert C. Zinke, "Cost-Benefit Analysis and Administrative Legitimation," *Policy Studies Journal* 6 (Autumn 1987): 63–88; and Steven Kelman, "Cost-Benefit Analysis: An Ethical Critique," in Glickman and Gough, *Readings in Risk*, 129–136.

50. This draws on the discussion of cost-benefit analysis in David L. Weimer and Aidan R. Vining, *Policy Analysis: Concepts and Practices* (Englewood Cliffs, N.J.: Prentice-Hall, 1989), 239–265.

51. Ibid., 265.

52. U.S. Environmental Protection Agency, *EPA's Use of Cost-Benefit Analysis, 1981–1986.*

53. In southern California, the ozone problem is so serious that the South Coast Air Quality Management District must clamp down on nearly all sources, no matter what the relative costs. See, for example, Larry B. Stammer, "AQMD Imposes Pollution Controls on Commercial Bakeries," *Los Angeles Times*, January 5, 1991, B1. The twenty-four large commercial bakeries affected emit about 4.1 tons of VOCs a day, almost what a medium-sized oil refinery emits. The controls would add less than a penny to the price of a one-pound loaf of bread.

54. See, for example, Joseph C. Reinert, Susan G. Slotnick, and Donn J. Viviani, "A Discussion of the Methodologies Used in Pesticide Risk-Benefit Analysis," *Environmental Professional* 12 (1990): 94–100.

55. Zinke, "Cost-Benefit Analysis and Administrative Legitimation," 81.

56. Michael Gough, "How Much Cancer Can EPA Regulate Away?" *Risk Analysis* 10, no. 1 (1990): 1–6.

57. Ibid., 1.

58. Brian E. Henderson, Ronald K. Ross, and Malcolm C. Pike, "Toward the Primary Prevention of Cancer," *Science*, no. 254 (November 22, 1991): 1131–1138, on 1137.

59. An example is discussed in Phil Brown, "Popular Epidemiology: Community Response to Toxic-Waste Induced Disease in Woburn, Massachusetts," *Science, Technology, and Human Values* 12 (Summer/Fall 1987): 78–85.

60. U.S. Environmental Protection Agency, *Reducing Risk, Report of the Ecology and Welfare Subcommittee* (Washington, D.C., September 1990), 30–33. There also is a very strong stream of criticism of conventional welfare economics and the way it treats ecological resources. See Herman E. Daly and John B. Cobb, Jr., *For the Common Good: Redirecting the Economy Toward Community, the Environment, and a Sustainable Future* (Boston: Beacon Press,

1989). Their critique of discounting is on pp. 151–158. They observe, "Discounting is a messy and disputed business about which economists themselves disagree" (152).

61. Peter Passell, "Rebel Economists Add Ecological Cost to Price of Progress," *New York Times,* November 27, 1990, C1, C13.

62. Cairncross, *Costing the Earth*, 32.

63. U.S. Environmental Protection Agency, *Reducing Risk, Report of the Ecology and Welfare Subcommittee,* 30.

64. Douglas E. MacLean, "Comparing Values in Environmental Policies: Moral Issues and Moral Arguments," in Hammond and Coppock, *Valuing Health Risks, Costs, and Benefits for Environmental Decision-Making,* 104.

CHAPTER FIVE: PROBLEMS

1. These arguments are at the core of modern American political science. The principal analysis of interest group liberalism is Lowi, *The End of Liberalism*. The classic work on incrementalism is Lindblom, "The Science of Muddling Through."

2. For a good analysis of this test of political support for the environment in the early 1980s, see Walter A. Rosenbaum, "The Politics of Public Participation in Hazardous Waste Management," in James P. Lester and Ann O'M. Bowman, eds., *The Politics of Hazardous Waste Management* (Durham: Duke University Press, 1983).

3. Mitchell, "Public Opinion and the Green Lobby," 83.

4. These trends in strength and salience are drawn from the discussion in Mitchell, "Public Opinion and the Green Lobby," 83–87. For a slightly more recent analysis, see Christopher J. Bosso, "After the Movement: Environmental Activism in the 1990s," in Vig and Kraft, *Environmental Policy in the 1990s,* 31–50.

5. The concept of "focusing events" comes from Kingdon's *Agendas, Alternatives, and Public Policy*. It describes events that focus public attention on an issue or problem where public fears or concerns had existed but had not been brought to the surface of policy debates. Besides the Santa Barbara and *Exxon Valdez* oil spills, examples of focusing events are the Three-Mile Island nuclear accident, the Love Canal waste site (discussed later in the chapter), and the chemical accident in Bhopal, India, in 1984, in which thousands of people died.

6. See Caldwell, "Environment: A New Focus for Public Policy?"

7. Let me qualify this by saying that even for small, simple societies living in some kind of ecological balance with their surroundings, environmental problems are probably inescapable. They will be affected by the actions of others, whether it is deforestation, erosion of topsoil, salinization or pollution of waters they rely on, or other trends. Should any societies escape these calamities, remember that they may have to cope someday with changes in the climate induced by global warming.

8. A good discussion of environmental policy from a comparative perspec-

tive is Sheldon Kamieniecki and Eliz Sanasarian, "Conducting Comparative Research in Environmental Policy," *Natural Resources Journal* 30 (Spring 1990): 321–339.

9. Baruch Fischhoff, "Report from Poland: Science and Politics in the Midst of Environmental Disaster," *Environment* 33 (March 1991): 13.

10. V. M. Kotlyakov, "The Aral Sea Basin: A Critical Environmental Zone," *Environment* 33 (January/February 1991): 4–9, 36–38. A recent, excellent book on environmental problems and policies in the former Soviet Union is D. J. Peterson, *Troubled Lands: The Legacy of Soviet Environmental Destruction* (Boulder: Westview Press, 1993).

11. William K. Reilly, "Beyond 'Big Green': Has the Environmental Movement Crested?" Address to the Commonwealth Club of California, San Francisco, January 9, 1991.

12. See Richard J. Tobin, "Environment, Population, and Development in the Third World," in Vig and Kraft, *Environmental Policy in the 1990s,* 279–300; and John W. Mellor, "The Intertwining of Environmental Problems and Poverty," *Environment* 30 (November 1988): 8–13, 28–30.

13. This discussion of international problems is adapted from Fiorino, "Institutional Responses to the Emerging Global Environment," 177–192. For a comprehensive analysis of international environmental issues and institutions, see Caldwell, *International Environmental Policy.* A useful set of essays is Jessica Tuchman Mathews, ed., *Preserving the Global Environment: The Challenge of Shared Leadership* (New York: W. W. Norton, 1991).

14. This draws on Giandomenico Majone, "The International Dimension," in Harry Otway and Malcolm Petu, *Regulating Industrial Risks: Science, Hazards, and Public Protection* (London: Butterworths, 1985), 40–56.

15. National Academy of Sciences, *Policy Implications of Greenhouse Warming* (Washington, D.C.: National Academy Press, 1991), 25.

16. U.S. Environmental Protection Agency, *Unfinished Business, Overview Report* (Washington, D.C., February 1987).

17. U.S. Environmental Protection Agency, *Reducing Risk.*

18. See Thomas E. Waddell and Blair T. Bower, *Managing Agricultural Chemicals in the Environment: The Case for a Multimedia Approach* (Washington, D.C.: Conservation Foundation, 1988).

19. U.S. Environmental Protection Agency, *Citizen's Guide to Radon,* 2d ed. (Washington, D.C., May 1992). Also see the "Radon Case Study" in U.S. Environmental Protection Agency, *Reducing Risk, Report of the Human Health Subcommittee* (Washington, D.C., September 1990).

20. Cited in *Citizen's Guide to Radon,* 2d ed. EPA's earlier estimates were 21,600 annual deaths as most likely within a range of 8,400 to 43,200. A reassessment of the risk analyses led the agency to reduce its estimates by about 30 percent to the number cited in the *Citizen's Guide.*

21. Abelson, "Radon Today," 95.

22. This is related in Sheldon Krimsky and Alonzo Plough, *Environmental Hazards: Communicating Risks as a Social Process* (Dover, Mass.: Aubern House, 1988), 132–134. Their chap. 4 is an excellent overview on radon.

23. An excellent resource on groundwater, and the source of many of

the figures given here, is Conservation Foundation, *Groundwater Protection* (Washington, D.C.: Conservation Foundation, 1987).

24. From the Phase II Report of EPA's National Survey of Pesticides in Drinking Water Wells, p. 4.

25. U.S. Environmental Protection Agency, *Protecting the Nation's Ground Water: EPA's Strategy for the 1990s* (Washington, D.C., July 1991).

26. The discussion of lead draws on U.S. Environmental Protection Agency, Office of Policy, Planning, and Evaluation, *Background Paper: The Lead Problem and Lead Management Activities* (Washington, D.C., August 1990); *OECD Cooperative Risk Reduction Activities for Lead* (Paris: Organization for Economic Cooperation and Development, Draft, 1991).

27. For a discussion of the many kinds of health risks from lead, see Ellen K. Silbergeld, "Lead in the Environment: Coming to Grips with Multisource Risks and Multifactorial Endpoints," *Risk Analysis* 9 (June 1989): 137–140.

28. See U.S. Environmental Protection Agency, *Environmental Equity,* which concluded that "a significantly higher percentage of Black children compared to White children have unacceptably high blood lead levels" (Vol. 1, p. 11).

29. This discussion draws on U.S. Environmental Protection Agency, Office of Policy, Planning, and Evaluation, *Threats to Biological Diversity in the United States* (Washington, D.C., September 1990). Two other useful but general reports are Office of Technology Assessment, *Technologies to Maintain Biological Diversity* (Washington, D.C., 1987); and W. V. Reid and K. R. Miller, *Keeping Options Alive—The Scientific Basis for Conserving Biodiversity* (Washington, D.C.: World Resources Institute, 1989).

30. U.S. Environmental Protection Agency, *Threats to Biological Diversity in the United States*, 24.

31. A good discussion is Alexander Leaf, "Potential Health Effects of Global Climatic and Environmental Changes," *New England Journal of Medicine,* no. 321 (December 7, 1989): 1577–1583.

32. U.S. Environmental Protection Agency, *Reducing Risk, Report of the Ecology and Welfare Subcommittee,* 27.

33. Kingdon, *Agendas, Alternatives, and Public Policies,* 3. The next few paragraphs draw on Kingdon's analysis.

34. On why genetic engineering has not risen further on the political (as opposed to the narrowly scientific) agenda, see Amal Kawar, "Issue Definition, Democratic Participation, and Genetic Engineering," *Policy Studies Journal* 17 (Summer 1989): 719–744.

35. Christopher J. Bosso, *Pesticides and Politics: The Life Cycle of a Public Issue* (Pittsburgh: University of Pittsburgh Press, 1987), 16.

36. Landy, Roberts, and Thomas, *Asking the Wrong Questions*, 91.

37. Ibid., 103.

38. Natural Resources Defense Council, *Intolerable Risk: Pesticides in Our Children's Food* (New York: Natural Resources Defense Council, 1989). See John A. Moore, "Speaking of Data: The Alar Controversy," *EPA Journal* 15 (May/June 1989): 5–9, for an EPA official's point of view. Moore was the deputy administrator of EPA at the time.

39. Examples are Eric Schmitt, "Miles of L.I. Beaches Are Closed by a

Wave of Sewage and Debris," *New York Times,* July 7, 1988, A1; Fox Butterfield, "To Swim or Not to Swim as Needles Wash Ashore," *New York Times,* July 8, 1988, B4; Jesus Rangel, "Waste Drowns Summer Along the Jersey Shore," *New York Times,* July 29, 1988, B2; Sally Squires, "Needles on the Beach: The Growing Problem of Medical Waste," *Washington Post Health,* August 23, 1988, 12–14.

40. The Medical Waste Tracking Act of 1988 was added as a new Subtitle J to the Solid Waste Disposal Act, 42 U.S.C. Sections 6901 et seq.

41. A good analysis is Kathryn D. Wagner, "Medical Wastes on Our Coasts: Issues and Impacts," paper presented at the Second International Conference on Marine Debris, Honolulu, Hawaii (April 2–7, 1989).

42. Ibid., 13.

43. This section draws on Helen Ingram and Carole L. Mintzer, "How Atmospheric Research Changed the Political Climate," in *Global Climate Change: The Meeting of Science and Policy* (Tucson: Udall Center for Studies in Public Policy, University of Arizona, March 1990), 1–11.

44. Ibid., 3.

45. An example on the cleanup side of the hazardous waste issue is Milton Russell, "Wasteful Waste Disposal?" *Washington Post,* March 20, 1992, A25.

46. The release of the National Academy of Sciences report, *Pesticides in the Diets of Infants and Children,* stimulated greater interest in the dietary risks of pesticides in the summer and fall of 1993.

47. William K. Stevens, "E.P.A. Acts to Reshuffle Environment Priorities," *New York Times,* January 26, 1991, 11.

48. U.S. Environmental Protection Agency, Science Advisory Board, *Unfinished Business: A Comparative Assessment of Environmental Problems* (Washington, D.C., February 1987). The five-volume report consists of an overview and four appendixes. The latter cover work group reports on cancer, noncancer health, ecological, and welfare effects. The comparative risk project is discussed in more detail in Frederick W. Allen, "Towards a Holistic Appreciation of Risk: The Challenge for Communicators and Policymakers," *Science, Technology, and Human Values* 12 (Summer/Fall 1987): 138–143; and in Richard Morgenstern and Stuart Sessions, "Weighing Environmental Risks: EPA's Unfinished Business," *Environment* 30 (July/August 1988): 15–17, 34–39.

49. The comparative risk projects in turn were preceded by a series of integrated environmental management projects, which are not discussed here. They are examined in Steven Cohen and Gary Weiskopf, "Beyond Incrementalism: Cross-Media Environmental Management in the Environmental Protection Agency," in Michael S. Hamilton, ed., *Regulatory Federalism, Natural Resources, and Environmental Management,* 47–63.

50. The project and report are discussed in more detail in Daniel J. Fiorino, "Can Problems Shape Priorities? The Case of Risk-Based Environmental Planning," *Public Administration Review* 50 (January/February 1990): 82–90; Allen, "Towards a Holistic Appreciation of Risk"; and Morgenstern and Sessions, "Weighing Environmental Risks."

51. Discussed in U.S. Environmental Protection Agency, *Unfinished Business, Overview Report,* 91–94. The report's conclusions about public attitudes were based on data collected by the Roper Organization.

52. Fiorino, "Can Problems Shape Priorities?" 84–85. On the first three regional projects, see U.S. Environmental Protection Agency, Office of Policy, Planning, and Evaluation, *Comparing Risks and Setting Priorities: Overview of Three Regional Projects* (Washington, D.C., August 1989).

53. The states that have completed rankings include Washington, Colorado, Vermont, Louisiana, and Michigan. Maine, Alabama, Oregon, Utah, California, Hawaii, Texas, Arizona, and Illinois had projects under way as of the end of 1993. The eight cities are Atlanta; Seattle; Jackson, Mississippi; Cleveland; Columbus, Ohio; Houston; Charlottesville, Virginia; and Elizabeth River, Virginia. Information on these projects can be obtained from the Regional and State Planning Branch, Office of Strategic Planning and Environmental Data, at EPA's Washington headquarters.

54. Jones, *An Introduction to the Study of Public Policy,* 3d ed., 64.

CHAPTER SIX: STRATEGIES

1. Grover Starling, *Strategies for Policy Making* (Chicago: Dorsey Press, 1988), 219.

2. Richard F. Elmore, "Instruments and Strategy in Public Policy," *Policy Studies Review* 7 (Autumn 1987): 174–186, on 184. On related issues, see Patricia W. Ingraham, "Toward More Systematic Consideration of Policy Design," *Policy Studies Journal* 15 (June 1987): 611–628.

3. Elmore, "Instruments and Strategy in Public Policy," 180.

4. Ibid., 175.

5. On labeling as a way to reduce risks, see Susan G. Hadden, *Read the Label: Reducing Risk by Providing Information* (Boulder: Westview Press, 1986).

6. Work on risk communication has grown as rapidly as the need for better ways of influencing behavior on radon and other problems. Good starting points are Krimsky and Plough, *Environmental Hazards*; and National Academy of Sciences, *Improving Risk Communication* (Washington, D.C.: National Academy Press, 1989).

7. U.S. Environmental Protection Agency, *Citizen's Guide to Radon,* 2d ed. Some states are supplementing risk communication with requirements for testing and notice in real estate transactions.

8. Alan Carlin, *The United States Experience with Economic Incentives to Control Environmental Pollution* (Washington, D.C.: U.S. Environmental Protection Agency, July 1992), 6-5. This refers to the study by Michael S. Baram, Patricia S. Dillon, and Betsy Ruffle, *Managing Chemical Risks: Corporate Response to SARA Title III* (Boston: Tufts University, Center for Environmental Management, 1990).

9. Hadden, *A Citizen's Right to Know.*

10. Described in Melinda Haag, "Proposition 65's Right-to-Know Provision: Can It Keep Its Promise to California Voters?" *Ecology Law Quarterly* 14 (1987): 685–712.

11. Section 25249.6. Quoted in Mazmanian and Morell, *Beyond Superfailure,* 171. An overview of Proposition 65 and "toxics populism" appears on pp. 169–174.

12. The illustrations do not cover the waste and drinking water programs, both of which rely almost entirely on direct regulation that is supplemented with some (but probably too little) risk communication and technical assistance.

13. Peter Huber, "The Old-New Division in Risk Regulation," *Virginia Law Review* 89 (1983): 1025–1107, on 1026.

14. U.S. Environmental Protection Agency, *Reducing Risk, Report of the Strategic Options Subcommittee* (Washington, D.C., September 1990), Appendix C, 33.

15. On weaknesses of direct regulation, see Richard B. Stewart, "Controlling Environmental Risks Through Economic Incentives," *Columbia Journal of Environmental Law* 13 (1988): 153–169; Michael H. Levin and Barry S. Elman, "The Case for Economic Incentives," *Environmental Forum* (January/February 1990): 7–11; Bruce A. Ackerman and Richard B. Stewart, "Reforming Environmental Law: The Democratic Case for Market Incentives," *Columbia Journal of Environmental Law* 13 (1988): 171–199.

16. Kneese and Schulze, *Pollution, Prices, and Public Policy.*

17. For a discussion, see Robert W. Hahn and Robert N. Stavins, "Incentive-Based Regulation: A New Era from an Old Idea?" (Harvard University, Energy and Environmental Policy Center [E-90-13]), 21–38. Examples of the interest in incentive approaches are two reports issued by Senators Timothy Wirth and John Heinz: *Project 88, Harnessing Market Incentives to Protect the Environment: Initiatives for the New President* (Washington, D.C., 1988); and *Project 88—Round II: Incentives for Action: Designing Market-Based Incentive Strategies* (Washington, D.C., 1991).

18. Kneese and Schulze, *Pollution, Prices, and Public Policy,* 6.

19. This section draws at several points on Carlin, *The United States Experience with Economic Incentives,* 1–2. Another excellent source is Organization for Economic Cooperation and Development, *Environmental Policy: How to Apply Economic Instruments* (Paris: Organization for Economic Cooperation and Development, 1991).

20. J. B. Opschoor and Hans P. Vos, *Economic Instruments for Environmental Protection* (Paris: Organization for Economic Cooperation and Development, 1989).

21. U.S. Environmental Protection Agency, Office of Policy, Planning, and Evaluation, *Economic Incentives: Options for Environmental Protection* (Washington, D.C., March 1991), chap. 4.

22. Carlin, *The United States Experience with Economic Incentives,* 9-4.

23. Kneese and Schulze, *Pollution, Prices, and Public Policy,* 99–101.

24. U.S. Environmental Protection Agency, *Economic Incentives,* 5-5 to 5-9. The 1990 CAAA adopted a fee on "excess" emissions of VOCs in certain nonattainment areas if they fail to attain the ozone NAAQS by their deadline. Fees were set in the statute at $5,000 per ton, to be adjusted for inflation.

25. Claudia Copeland, *Funding Water Quality Programs: Revenues for a National Clean Water Investment Corporation* (Washington, D.C.: Congressional Research Service, July 1992). This same report examined the use of fees on pesticide active ingredients and fertilizers—fees applied to inputs. Both fees would fund a national corporation for improving water quality.

26. On the bubble and emissions trading programs, see Richard A. Liroff,

Reforming Air Pollution Regulation: The Toil and Trouble of EPA's Bubble (Washington, D.C.: Conservation Foundation, 1986); Thomas H. Tietenberg, *Emissions Trading: An Exercise in Reforming Pollution Policy* (Washington, D.C.: Resources for the Future, 1985); Michael H. Levin, "Getting There: Implementing the Bubble Policy," in Eugene Bardach and Robert A. Kagan, eds., *Social Regulation: Strategies for Reform* (San Francisco: Institute for Contemporary Studies, 1982), 59–92; Errol Meidinger, "The Development of Emissions Trading in U.S. Air Pollution Regulation," in Keith Hawkins and John M. Thomas, eds., *Making Regulatory Policy* (Pittsburgh: University of Pittsburgh Press, 1989), 153–194.

27. Ed Reynolds, "The Acid Test on Acid Rain," *Environmental Protection* (October 1990): 36–41, on 38.

28. Carlin, *The United States Experience with Economic Incentives,* 5–6. This estimate is from EPA's Regulatory Impact Analysis for the acid rain trading rules. Others have estimated even higher savings.

29. Ibid., 5–8.

30. These and other marketable permit approaches are discussed in Carlin, *The United States Experience with Economic Incentives,* chap. 5.

31. U.S. Environmental Protection Agency, *Economic Incentives,* 4-15 to 4-17.

32. See the discussion, "Agriculture and the Environment in a Changing World Economy," in Conservation Foundation, *State of the Environment,* 341–406.

33. These studies are summarized in Carlin, *The United States Experience with Economic Incentives,* 2-2 to 2-6. The other estimates for savings from trading programs are from chap. 5 of Carlin's book.

34. The quote is from "EPA Staff Warn of Equity Concerns in Emissions Trading Programs," *Inside EPA* 14 (July 2, 1993): 16.

35. U.S. Environmental Protection Agency, Office of Policy, Planning, and Evaluation, Regulatory Innovations Staff, *Draft Guidelines for the Application of Emissions Trading to Air Pollution Control* (Washington, D.C., April 1990), 3.

36. Described in U.S. Environmental Protection Agency, *Risk Assessment and Risk Management: A Framework for Decision Making* (Washington, D.C., 1984).

37. National Research Council, *Risk Assessment in the Federal Government: Managing the Process* (Washington, D.C.: National Academy Press, 1983). On EPA's orientation toward risk, see William D. Ruckelshaus, "Risk, Science, and Democracy," *Issues in Science and Technology* 1 (Spring 1985): 19–38; Milton Russell, "Environmental Protection: Laying the Groundwork for the Year 2000," *Environmental Forum* 10 (1986): 7–12; and Russell, "Environmental Protection for the 1990s and Beyond." An introduction to risk management is Conservation Foundation, *Risk Assessment and Risk Control* (Washington, D.C.: Conservation Foundation, 1985).

38. See U.S. Environmental Protection Agency, *Pollution Prevention Strategy* (Washington, D.C., January 1991). Prevention as a strategy has been endorsed by federal and state legislatures. Congress passed the Pollution Prevention Act in 1990. Since 1987, twenty-seven states have enacted laws

promoting prevention as the preferred method for managing waste. Some are described in EPA's *Pollution Prevention News* (July/August 1992).

39. The Asbestos School Hazard Abatement Act of 1984 (P.L. 98-377) provided loans and grants for asbestos abatement, but removal was the only abatement method funded. The Asbestos Hazard Emergency Act of 1986 (P.L. 99-519) required EPA to issue rules for assessing asbestos-containing materials in schools.

40. These very issues were faced in developing U.S. Environmental Protection Agency, *Pesticides-and-Groundwater Strategy* (Washington, D.C., October 31, 1991).

41. For discussions, see the series of studies on environmental integration by the Conservation Foundation, including Haigh and Irwin, eds., *Integrated Pollution Control in Europe and North America*; Rabe, *Fragmentation and Integration in State Environmental Management*; and Frances H. Irwin, Edwin H. Clark, and J. Clarence Davies, *Controlling Cross-Media Pollutants* (Washington, D.C.: Conservation Foundation, 1984).

42. U.S. Environmental Protection Agency, *Reducing Risk*, 23.

43. Frances H. Irwin, "An Integrated Framework for Preventing Pollution and Protecting the Environment" (Washington, D.C.: Conservation Foundation, Draft, April 1991), 17. A related discussion is Robert W. Hahn and Eric H. Males, "Can Regulatory Institutions Cope with Cross Media Pollution?" *Journal of the Air and Waste Management Association* 40 (January 1990): 24–31.

44. Discussed in Terry Davies, "The United States: Experimentation and Fragmentation," in Haigh and Irwin, *Integrated Pollution Control in Europe and North America,* 51–66.

45. On the Amoco project, see Mahesh K. Podar, Howard Klee, Jr., and Robert Kerr, "Integrated Environmental Management: A Case Study of Pollution Prevention at a Refinery," paper presented at the annual meeting of the Air and Waste Management Association, June 1991. For an interesting account of the effort to integrate policy making in the United Kingdom, see Susan Owens, "The Unified Pollution Inspectorate and Best Practicable Environmental Option in the United Kingdom," in Haigh and Irwin, *Integrated Pollution Control in Europe and North America*, 169–208.

46. U.S. Environmental Protection Agency, *Strategy for Reducing Lead Exposures* (Washington, D.C., February 21, 1991).

47. As of late 1993, some but not all of the elements of the lead strategy had been carried out, and the future of some of these elements is uncertain. My point is to use the lead strategy as a case for discussing instruments and strategies in environmental policy making. The lead strategy itself will continue to evolve.

CHAPTER SEVEN: PROSPECTS

1. Of course, unachievable deadlines can perform political functions for environmental advocates. In his "Pollution Deadlines and the Coalition for Failure," *Public Interest* (Spring 1984): 123–134, R. Shep Melnick asserts that

"perpetual failure to meet deadlines gives those advocating more pollution control a key rhetorical advantage" (131). Not meeting deadlines reminds us how much is left to be done.

2. Most of these are summarized in U.S. Environmental Protection Agency, *Meeting the Environmental Challenge: EPA's Review of Progress and New Directions in Environmental Protection* (Washington, D.C., December 1990). A good discussion of trends is Conservation Foundation, *State of the Environment*.

3. Cited in Keith Bradsher, "Lower Pollution Tied to Prosperity," *New York Times*, October 28, 1991, D3. Another study comparing economic growth and environmental policy in the American states found that states having strong environmental programs also had healthier economies. Although it is difficult to draw clear lessons from the study, it at least supports a conclusion that growth is not incompatible with environmental protection. See Stephen M. Meyer, *Environmentalism and Economic Prosperity: Testing the Environmental Impact Hypothesis* (Cambridge: Massachusetts Institute of Technology, Project on Environmental Politics and Policy, October 5, 1992).

4. World Commission on Environment and Development, *Our Common Future* (London: Oxford University Press, 1987), 43. The commission was chaired by Gro Brundtland, former prime minister of Norway. Offering even a general definition of the concept of sustainable development is risky, because it means so many things to so many people. "Every environmentally aware politician is in favor of it," Frances Cairncross observes, "a sure sign that they do not understand what it means" (*Costing the Earth,* 26).

5. Francis Fukuyama, *The End of History and the Last Man* (New York: Avon Books, 1992), 114.

6. Ibid., 114–115.

7. For a comparative look at this issue, see Dorothy Nelkin, *Technological Decisions and Democracy: European Experiments in Public Participation* (Beverly Hills, Calif.: Sage, 1977).

8. This draws on the discussion in Fiorino, "Citizen Participation and Environmental Risk: A Survey of Institutional Mechanisms."

9. Ortwin Renn, Thomas Webler, and Branden B. Johnson, "Public Participation in Hazardous Waste Management: The Use of Citizens Panels in the U.S.," *Risk—Issues in Health and Safety* 2 (Summer 1991): 197–226. On the citizen's panel generally, see N. Crosby, J. M. Kelly, and P. Schaefer, "Citizens Review Panels: A New Approach to Citizen Participation," *Public Administration Review* 46 (1986): 170–178.

10. For a description, see U.S. Environmental Protection Agency, *Rationalizing Regulation: The EPA Clusters Project* (Washington, D.C., 1991).

11. Discussed in Davies, "The United States: Experimentation and Fragmentation."

12. Owens, "The Unified Pollution Inspectorate and Best Practicable Environmental Option in the United Kingdom," 172.

13. Ibid., 177.

14. This grasp is evident in Al Gore, *Earth in the Balance: Ecology and the Human Spirit* (New York: Plume, 1992).

15. U.S. Environmental Protection Agency, *Pollution Prevention Strategy.*

16. Harry Freeman et al., "Industrial Pollution Prevention: A Critical Review," *Journal of the Air and Waste Management Association* 42 (May 1992): 629.

17. The examples are taken from ibid., 631, and from Scott McMurray, "Chemical Firms Find that It Pays to Reduce Pollution at the Source," *Wall Street Journal,* June 11, 1991, A1, A6.

18. "EPA Launches Program to Spur Innovative Corporate Strategies," *Inside EPA* 15 (April 22, 1994).

19. U.S. Environmental Protection Agency, *Economic Incentives,* 2-7. The discussion of volume-based pricing is on pp. 2-7 to 2-12.

20. Robert Repetto, Roger C. Dower, Robin Jenkins, and Jacqueline Geoghegan, *Green Fees: How a Tax Shift Can Work for the Environment and the Economy* (Washington, D.C.: World Resources Institute, 1992). Fees were also the centerpiece of the chapter on the environment in the report from the Progressive Policy Institute: Robert Stavins and Thomas Grumbly, "The Greening of the Market: Making the Polluter Pay," in Will Marshall and Martin Schram, eds., *Mandate for Change* (New York: Berkley Books, 1993), 197–216.

21. From the U.S. Environmental Protection Agency's strategic plan, *EPA—Preserving Our Future Today* (Washington, D.C., Draft 3A, November 29, 1991), 10.

22. These examples are taken from *Environmental Indicators: Policies, Programs, and Success Stories,* prepared by the Environmental Results and Forecasting Branch for an Environmental Indicators Workshop, July 17–19, 1991.

23. Described in U.S. Environmental Protection Agency, *Setting National Goals for Environmental Protection* (Washington, D.C., Public review document, April 1994).

24. Robert D. Bullard, *Dumping in Dixie: Race, Class, and Environmental Quality* (Boulder: Westview Press, 1990).

25. For a review of these studies, see Joyce A. Baugh, "African-Americans and the Environment: A Review Essay," *Policy Studies Journal* 19 (Spring 1991): 182–191. An excellent collection of articles can be found in Robert D. Bullard, ed., *Confronting Environmental Racism: Voices from the Grassroots* (Boston: South End Press, 1993). Also see Richard J. Lazarus, "Pursuing 'Environmental Justice': The Distributional Effects of Environmental Protection," *Northwestern University Law Review* 87 (Spring 1993): 787–857.

26. Baugh, "African-Americans and the Environment," 184.

27. Summarized in U.S. Environmental Protection Agency, *Environmental Equity:* Vol. 1, *Work Group Report to the Administrator.*

28. "Federal Actions to Address Environmental Justice in Minority Populations and Low-Income Populations" (Office of the White House Press Secretary, February 11, 1994), 1. See John H. Cushman, Jr., "Clinton to Order Effort to Make Pollution Fairer," *Washington Post,* February 10, 1994, 1.

29. Of course, a time preference could go the other way in the process of valuing ecological resources. Some people have argued for a "negative" discount rate that attaches greater value to the benefits derived from

ecological resources in the future. With a negative discount rate, the dollar value of a wetland in 100 years would be far greater than the value of that resource now.

30. Brian Norton has observed that on "questions regarding the fairness of intergenerational distribution of risk . . . it is difficult to see how the standard economic model can adequately conceptualize these essentially ethical considerations." Brian G. Norton, *Toward Unity Among Environmentalists* (New York: Oxford University Press, 1991), 210.

31. See Roberto Suro, "Pollution-Weary Minorities Try Civil Rights Tack," *New York Times,* January 11, 1993, A1.

32. In addition to Daly and Cobb's *For the Common Good,* there are EPA's *Reducing Risk,* the report of the National Commission on the Environment, and growing lists of publications by economists, philosophers, and others. An example is Norton, *Toward Unity Among Environmentalists.*

Glossary

ADMINISTRATIVE PROCEDURE ACT (APA). The U.S. statute, passed in 1946, that sets out the legal process and framework by which federal administrative agencies issue regulations.

AMBIENT POLLUTION STANDARDS. Goals or limits on concentrations of pollutants in the surrounding air or water.

ASSISTANT ADMINISTRATORS (AAS). The senior officials in the U.S. Environmental Protection Agency who direct the ten or so major EPA offices in Washington. Because they are political (presidential) appointees, they almost always come and go with changes in the presidential administrations.

BOUNDED RATIONALITY. The notion that although people aspire to be rational in organizations, they must make decisions in the face of limited time, resources, information, and human capacities. It is more realistic for people to aspire to a limited but more realistic bounded rationality than an ideal of complete rationality.

COMPARATIVE RISK PROJECTS (RISK-BASED PLANNING). A series of projects sponsored by EPA and state and local agencies in which different sources of environmental risk are identified, assessed, and ranked, as a basis for devising an environmental management strategy. They involve substantial community participation.

CONGRESSIONAL OVERSIGHT COMMITTEES. The committees within the U.S. Senate and House of Representatives that monitor the implementation of the laws by federal agencies, including environmental agencies. The Senate Committee on Environment and Public Works, for example, is the most important oversight committee for environmental policy in the Senate. Committees usually work through more specialized subcommittees.

CONVENTIONAL POLLUTANTS. Common and persistent pollutants that are harmful over time and in quantity but not immediately dangerous at typical levels in the ambient air or water. Examples are sulfur oxide and particulates, two common forms of air pollution.

COST-BENEFIT ANALYSIS. A set of methods for assigning dollar value to the costs (undesirable or negative effects) and to the benefits (desirable or positive effects) of policies and comparing them.

COST-EFFECTIVENESS ANALYSIS. A method for comparing the results per unit of expenditure of dollars for environmental controls. It enables us to compare the costs of a given measure per unit of pollution reduced, lives saved, illnesses avoided, or other result.

DISCOUNTING. A way of accounting for costs and benefits that are incurred or realized over time. Analysts discount the future costs and sometimes the benefits of government policies to their present value to make them comparable.

EFFLUENT GUIDELINES. National, uniform, technology-based limits on discharges from industrial sources of water pollution. These are similar to NSPS in the air program and are based on best available technology, with some consideration given to cost and feasibility.

EMISSION STANDARDS. Limits on the amount of a pollutant that may be emitted from a fixed source, such as a smokestack, automobile tailpipe, or water discharge pipe. Limits on water polluters are usually described as "effluent" standards; limits on air polluters are usually described as "emission" standards.

ENVIRONMENTAL GROUPS. The term used to describe the several public interest organizations created to advocate environmental values at various levels of government. Many of the national groups—such as the Sierra Club, Environmental Defense Fund, and National Wildlife Federation—are very influential in setting national policy on the environment and natural resources. Other groups are organized at the local level and focus on more specific, local issues.

ENVIRONMENTAL INDICATORS. Measures that can be used to assess status and trends in environmental quality.

ENVIRONMENTAL JUSTICE. Also known as "environmental equity," this refers to the issue of the distribution of environmental damages in society, particularly the disproportionate effects on minority and low-income groups. Environmental justice advocates argue that the burdens of pollution in American society fall more and to an unfair degree on minority and low-income groups.

ENVIRONMENTAL MEDIUM. Usually a term that refers to the pathway through which people or ecological resources are exposed to harm. Examples are drinking water, waste, air, and pesticides. Most U.S. environmental laws and programs are organized on the basis of the environmental medium of exposure.

FEDERALISM. A term for describing the sharing of power among levels of government. Authority for environmental policy is divided among federal, state, and local levels in the United States.

GREENHOUSE GASES. Air emissions that are generally acknowledged as contributing to the phenomenon of global warming (the greenhouse effect). Carbon dioxide is the most significant in terms of the quantity of emissions.

INCREMENTALISM. An approach to decision making in which decisions are made in small steps, at the margins of choice, with limited change from year to year. It is presented as more realistic and desirable than a fully "rational" approach to making decisions.

MARGINAL COSTS. The costs that are incurred as we move from one level of pollution control to another, usually more stringent one. Marginal costs almost always increase (on a per unit basis) as controls become stricter.

MARKETABLE PERMITS. A means of controlling pollution in which policy makers set a total level of emissions or discharges into the environment, allocate permits to allow sources to emit to that level, and then allow some form of trading of those permits.

MULTILATERAL DEVELOPMENT BANKS. International lending institutions (such as the World Bank or the Asian Development Bank) established to promote economic growth and development. They have been criticized in the past for not being sensitive enough to the environmental effects of their loan policies.

NATIONAL AMBIENT AIR QUALITY STANDARDS (NAAQS). A judgment by EPA as to what concentrations of several common air pollutants can exist in the ambient air and not harm public health, with an adequate margin of safety for sensitive groups in the population.

NEW SOURCE PERFORMANCE STANDARDS (NSPS). National, uniform limits on air emissions from categories of industrial sources. These reflect judgments as to what is the best technology available and affordable in a given industry category.

NONPOINT SOURCE OF POLLUTION. Scattered sources of air or water pollution, such as pesticides and fertilizers that run off farmland into streams. These usually are more difficult to regulate than are "point" sources.

OFFICE OF INFORMATION AND REGULATORY AFFAIRS (OIRA). The unit within the Office of Management and Budget charged with overseeing regulatory policy making by federal agencies. It has traditionally been most concerned with the costs of regulation.

POINT SOURCES OF POLLUTION. Identifiable sources of air emissions and water effluents, such as smokestacks and water discharge pipes.

POLICY AGENDA. The list of subjects or problems to which government officials and others are giving attention.

POLICY INSTRUMENTS. A means for government to respond to problems, including environmental problems. Information, direct regulation, and market or economic incentives are some of the main categories of environmental policy instruments.

POLICY SECTOR. A concept for describing one of the principal areas of public policy. Environment is a sector, as are energy, trade, transportation, and agriculture.

POLLUTION FEES. Charges applied to sources of pollution that increase as the amount of pollution increases. They are designed to provide an economic incentive for sources to reduce the amount of their emissions or discharges.

POLLUTION PREVENTION. A strategy that stresses the need to reduce the amounts of pollution at the source (i.e., by getting at the root causes of pollution) rather than just through controls at the point of emission or discharge. It includes anything that increases efficiency in the use of raw materials, energy, water, and other resources or protects resources through conservation.

REGULATORY IMPACT ANALYSIS (RIA). A form of economic analysis that compares the costs of a regulatory proposal to the benefits. An RIA is usually performed for "major" regulations.

REGULATORY NEGOTIATION. A way of developing regulatory proposals in which the responsible agency convenes a group of representatives of interests that are affected by a regulation, forms them into a committee, and attempts to reach agreement with them on the content and often the language of the proposed regulation.

RISK. The possibility of suffering harm. Its two dimensions are (1) the probability or likelihood of harm, and (2) the severity of the harm.

RISK ASSESSMENT. The process of estimating the harm that will result from exposure to an environmental contaminant. The risks usually are expressed in quantitative terms.

RISK-BASED STANDARD. A pollution standard that is based on a judgment about the estimated harm of a contaminant. Standards are set at a level that is considered necessary to reduce or keep risk from emissions or discharges at an "acceptable" level.

RISK COMMUNICATION. The use of information to make people aware of sources of risk (such as radon in their homes) and usually to cause them to take some kind of preventive or protective action.

RISK PERCEPTION. The study of people's attitudes toward risks, in particular, their views regarding the acceptability of risk and their behavior in response to what they think is harmful.

SOCIAL COSTS. The value that society places on the goods and services that are lost when resources are diverted to other ends.

STATE IMPLEMENTATION PLANS (SIPS). Plans that states are required to prepare under the Clean Air Act to describe how they will meet the National Ambient Air Quality Standards.

TECHNOLOGY-BASED STANDARD. A pollution standard that is based on a judgment about the best technology that is available (and perhaps affordable or economically feasible) for controlling emissions or discharges of a contaminant.

TOXIC POLLUTANTS. Contaminants that may pose serious harm even in small quantities. Examples are lead, benzene, mercury, and dioxin.

UNITED NATIONS CONFERENCE ON THE ENVIRONMENT AND DEVELOPMENT (UNCED). An international conference held in Rio de Janeiro in June 1992, on the twentieth anniversary of the first UN environmental conference, held in Stockholm in 1972.

UNITED NATIONS ENVIRONMENT PROGRAMME (UNEP). The agency within the United Nations that has served as a catalyst and forum for many international efforts at solving environmental problems. It was formed by the United Nations General Assembly in 1972, an outgrowth of the UN Conference on the Human Environment in Stockholm.

UNREASONABLE RISK. A standard under the Toxic Substances Control Act under which EPA determines whether it should take action to reduce exposures to a chemical substance or mixture.

Selected Bibliography

The literature on environmental policy making is large and is growing steadily. Many readers may want to pursue this subject in more detail. Here is a list of suggested readings, drawn from the books and articles cited in this book.

GENERAL WORKS ON ENVIRONMENTAL POLICY IN THE UNITED STATES

Bryner, Gary C. 1987. *Bureaucratic Discretion: Law and Policy in Federal Regulatory Agencies*. New York: Pergamon Press.

Bullard, Robert D. 1990. *Dumping in Dixie: Race, Class, and Environmental Quality*. Boulder: Westview Press.

———, ed. 1993. *Confronting Environmental Racism: Voices from the Grassroots*. Boston: South End Press.

Caldwell, Lynton K. 1963. "Environment: A New Focus for Public Policy?" *Public Administration Review* 23 (September): 132–139.

Conservation Foundation. 1987. *State of the Environment: A View toward the Nineties*. Washington, D.C.: Conservation Foundation.

Davies, J. Clarence. 1970. *The Politics of Pollution*. New York: Pegasus.

Eisner, Marc Allen. 1993. *Regulatory Politics in Transition*. Baltimore: Johns Hopkins University Press.

Fiorino, Daniel J. 1990. "Can Problems Shape Priorities? The Case of Risk-based Environmental Planning." *Public Administration Review* 50 (January/February): 82–90.

Hamilton, Michael S., ed. 1990. *Regulatory Federalism, Natural Resources, and Environmental Management*. Washington, D.C.: American Society for Public Administration.

Jasanoff, Sheila. 1990. *The Fifth Branch: Science Advisors as Policy Makers*. Cambridge: Harvard University Press.

Landy, Marc K., Marc J. Roberts, and Stephen R. Thomas. 1990. *The Environmental Protection Agency: Asking the Wrong Questions*. New York: Oxford University Press.
Lazarus, Richard J. 1993. "Pursuing 'Environmental Justice': The Distributional Effects of Environmental Protection." *Northwestern University Law Review* 87 (Spring): 787–857.
Marcus, Alfred A. 1980. *Promise and Performance: Choosing and Implementing an Environmental Policy*. Westport, Conn.: Greenwood Press.
National Commission on the Environment. 1993. *Choosing a Sustainable Future: The Report of the National Commission on the Environment*. Washington, D.C.: Island Press.
Rabe, Barry G. 1986. *Fragmentation and Integration in State Environmental Programs*. Washington, D.C.: Conservation Foundation.
Smith, V. Kerry. 1984. *Environmental Policy under Reagan's Executive Order*. Chapel Hill: University of North Carolina Press.
Stewart, Richard B. 1975. "The Reformation of American Administrative Law." *Harvard Law Review* 88: 1667–1813.
Vig, Norman J., and Michael E. Kraft, eds. 1984. *Environmental Policy in the 1980s: Reagan's New Agenda*. Washington, D.C.: CQ Press.
———. 1990. *Environmental Policy in the 1990s*. Washington, D.C.: CQ Press.

COMPARATIVE AND INTERNATIONAL ENVIRONMENTAL POLICY

Brickman, Ronald, Sheila Jasanoff, and Thomas Ilgen. 1985. *Controlling Chemicals: The Politics of Regulation in Europe and the United States*. Ithaca: Cornell University Press.
Caldwell, Lynton Keith. 1984. *International Environmental Policy: Emergence and Dimensions*. Durham: Duke University Press.
Gore, Al. 1992. *Earth in the Balance: Ecology and the Human Spirit*. New York: Plume.
Haigh, Nigel, and Frances Irwin., eds. 1990. *Integrated Pollution Control in Europe and North America*. Washington, D.C.: Conservation Foundation.
Jasanoff, Sheila. 1986. *Risk Management and Political Culture: A Comparative Study of Science in the Policy Context*. Beverly Hills, Calif.: Sage.
Kamienicki, Sheldon, and Eliz Sanasarian. 1990. "Conducting Comparative Research in Environmental Policy." *Natural Resources Journal* 30 (Spring): 321–339.
Lundqvist, Lennart J. 1980. *The Hare and the Tortoise: Clean Air Policies in the United States and Sweden*. Ann Arbor: University of Michigan Press.
Mathews, Jessica Tuchman, ed. 1991. *Preserving the Global Environment: The Challenge of Shared Leadership*. New York: W. W. Norton.
Pearson, Charles S., ed. 1987. *Multinational Corporations, Environment, and the Third World*. Durham: Duke University Press.
Peterson, D. J. 1993. *Troubled Lands: The Legacy of Soviet Environmental Destruction*. Boulder: Westview Press.

Redclift, Michael. 1987. *Sustainable Development: Exploring the Contradictions*. London: Methuen.

Vogel, David. 1986. *National Styles of Regulation: Environmental Policy in Great Britain and the United States*. Ithaca: Cornell University Press.

World Commission on Environment and Development. 1987. *Our Common Future*. London: Oxford University Press.

Zaelke, Durwood, Paul Orbuch, and Robert F. Housman. 1993. *Trade and the Environment: Law, Economics, and Policy*. Washington, D.C.: Island Press.

RISK AND POLICY ANALYSIS

Cairncross, Frances. 1993. *Costing the Earth*. Boston: Harvard Business School Press.

Cohrssen, John J., and Vincent T. Covello. 1989. *Risk Analysis: A Guide to Principles and Methods for Analyzing Health and Environmental Risks*. Washington, D.C.: Council on Environmental Quality.

Cropper, Maureen L., and Wallace E. Oates. 1992. "Environmental Economics: A Survey." *Journal of Economic Literature* 30 (June): 675–740.

Crouch, Edmund A. C., and Richard Wilson. 1982. *Risk/Benefit Analysis*. Cambridge: Ballinger.

Daly, Herman E., and John M. Cobb, Jr. 1989. *For the Common Good: Redirecting the Economy Toward Community, the Environment, and a Sustainable Future*. Boston: Beacon Press.

Douglas, Mary E. 1985. *Risk Acceptability According to the Social Sciences*. Beverly Hills, Calif.: Sage.

Douglas, Mary, and Aaron Wildavsky. 1982. *Risk and Culture: An Essay on the Selection of Technological and Environmental Dangers*. Berkeley, Los Angeles, and London: University of California Press.

Fiorino, Daniel J. 1990. "Technical and Democratic Values in Risk Analysis." *Risk Analysis* 9 (September): 293–299.

Glickman, Theodore S., and Michael Gough, eds. 1990. *Readings in Risk*. Washington, D.C.: Resources for the Future.

Hahn, Robert W., and John A. Hird. 1991. "The Costs and Benefits of Regulation: A Review and Synthesis." *Yale Journal on Regulation* 8 (Winter): 233–278.

Keyes, Ralph. 1985. *Chancing It: Why We Take Risks*. Boston: Little, Brown.

Kneese, Allen V., and Charles L. Schultze. 1975. *Pollution, Prices, and Public Policy*. Washington, D.C.: Brookings Institution.

Melnick, R. Shep. 1990. "The Politics of Cost-Benefit Analysis." In P. Brett Hammond and Rob Coppock, eds., *Valuing Health Risks, Costs, and Benefits for Environmental Decision Making*, 23–54. Washington, D.C.: National Academy Press.

National Academy of Sciences. 1983. *Risk Assessment in the Federal Government: Managing the Process*. Washington, D.C.: National Academy Press.

Nichols, Albert L., and Richard J. Zeckhauser. 1986. "The Perils of Prudence: How Conservative Risk Assessments Distort Regulation." *Regulation* (November/December): 13–24.

Repetto, Robert, et al. 1992. *Green Fees: How a Tax Shift Can Work for the Environment and the Economy*. Washington, D.C.: World Resources Institute.

Sexton, Ken, and Yolanda Banks Anderson. 1993. "Equity and Environmental Health: Research Issues and Needs." *Toxicology and Industrial Health* 9, special issue (September/October).

Shrader-Frechette, K. S. 1985. *Risk Analysis and Scientific Method: Methodological and Ethical Problems with Evaluating Societal Hazards*. Boston: D. Reidel.

Slovic, Paul. 1987. "Perception of Risk." *Science*, no. 236: 280–285.

Travis, Curtis C., et al. 1987. "Cancer Risk Management: A Review of 132 Federal Regulatory Decisions." *Environmental Science and Technology* 21: 415–420.

U.S. Environmental Protection Agency. 1987. *Unfinished Business: A Comparative Assessment of Environmental Problems*. Washington, D.C.

———. 1990. *Environmental Investments: The Cost of a Clean Environment*. Washington, D.C.

———. 1992. *Environmental Equity: Reducing Risks for All Communities*. Washington, D.C.

Weimer, David L., and Aidan R. Vining. 1989. *Policy Analysis: Concepts and Practices*. Englewood Cliffs, N.J.: Prentice-Hall.

Whittemore, Alice. 1983. "Facts and Values in Risk Analysis for Environmental Toxicants." *Risk Analysis* 1: 23–33.

WORKS ON SPECIFIC ENVIRONMENTAL ISSUES OR PROBLEMS

Bosso, Christopher J. 1987. *Pesticides and Politics: The Life Cycle of a Public Issue*. Pittsburgh: University of Pittsburgh Press.

Cohen, Richard E. 1992. *Washington at Work: Back Rooms and Clean Air*. New York: Macmillan.

Conservation Foundation. 1987. *Groundwater Protection*. Washington, D.C.: Conservation Foundation.

Mazmanian, Daniel, and David Morell. 1992. *Beyond Superfailure: America's Toxics Policy for the 1990s*. Boulder: Westview Press.

National Academy of Sciences. 1993. *Pesticides in the Diets of Infants and Children*. Washington, D.C.: National Academy Press.

———. 1991. *Policy Implications of Greenhouse Warming*. Washington, D.C.: National Academy Press.

Portney, Paul R., ed. 1990. *Public Policies for Environmental Protection*. Washington, D.C.: Resources for the Future.

Reitze, Arnold W. 1991. "A Century of Air Pollution Control Law: What's Worked, What's Failed, and What Might Work." *Environmental Law* 21: 1549–1646.

Waxman, Henry A. 1991. "An Overview of the Clean Air Act Amendments of 1990." *Environmental Law* 21: 1721–1816.

CITIZEN PARTICIPATION AND RISK COMMUNICATION

Barber, Benjamin R. 1984. *Strong Democracy: Participatory Politics for a New Age*. Berkeley, Los Angeles, and London: University of California Press.

Fiorino, Daniel J. 1988. "Regulatory Negotiation as a Policy Process." *Public Administration Review* 48 (July/August): 764–772.

———. 1989. "Environmental Risk and Democratic Process: A Critical Review." *Columbia Journal of Environmental Law* 14: 501–547.

———. 1990. "Citizen Participation and Environmental Risk: A Survey of Institutional Mechanisms." *Science, Technology, and Human Values* 12 (Spring): 226–243.

Hadden, Susan G. 1986. *Read the Label: Reducing Risk by Providing Information*. Boulder: Westview Press.

———. 1989. *A Citizen's Right to Know: Risk Communication and Public Policy*. Boulder: Westview Press.

Krimsky, Sheldon, and Alonzo Plough. 1988. *Environmental Hazards: Communicating Risks as a Social Process*. Dover, Mass.: Auburn House.

National Academy of Sciences. 1989. *Improving Risk Communication*. Washington, D.C.: National Academy Press.

Nelkin, Dorothy. 1977. *Technological Decisions and Democracy: European Experiments in Public Participation*. Beverly Hills, Calif.: Sage.

Reich, Robert B. 1985. "Public Administration and Public Deliberation: An Interpretive Essay." *Yale Law Journal* 94: 1617–1641.

Renn, Ortwin, Thomas Webler, and Branden B. Johnson. 1991. "Public Participation in Hazardous Waste Management: The Use of Citizens Panels in the U.S." *Risk—Issues in Health and Safety* 2 (Summer): 197–226.

Ruckelshaus, William D. 1985. "Risk, Science, and Democracy." *Issues in Science and Technology* 1 (Spring): 19–38.

STRATEGIES AND POLICY INSTRUMENTS

Carlin, Alan. 1992. *The United States Experience with Economic Incentives*. Washington, D.C.: U.S. Environmental Protection Agency.

Cook, Brian J. 1988. *Bureaucratic Politics and Regulatory Reform: The EPA and Emissions Trading*. New York: Greenwood Press.

Elmore, Richard F. 1987. "Instruments and Strategy in Public Policy." *Policy Studies Review* 7 (Autumn): 174–186.

Freeman, Harry, et al. 1992. "Industrial Pollution Prevention: A Critical Review," *Journal of the Air and Waste Management Association* 42 (May): 617–656.

Huber, Peter L. 1983. "The Old-New Division in Risk Regulation." *Virginia Law Review* 89: 1025–1107.

Liroff, Richard A. 1986. *Reforming Air Pollution Regulation: The Toil and Trouble of EPA's Bubble*. Washington, D.C.: Conservation Foundation.

Meidinger, Errol. 1989. "The Development of Emissions Trading in U.S. Air

Pollution Regulation." In Keith Hawkins and John M. Thomas, eds., *Making Regulatory Policy*, 153–194. Pittsburgh: University of Pittsburgh Press.

Opschoor, J. B., and Hans P. Vos. 1989. *Economic Instruments for Environmental Protection*. Paris: Organization for Economic Cooperation and Development.

Organization for Economic Cooperation and Development. 1991. *Environmental Policy: How to Apply Economic Instruments*. Paris: Organization for Economic Cooperation and Development.

Smart, Bruce. 1992. *Beyond Compliance: A New Industry View of the Environment*. Washington, D.C.: World Resources Institute.

Starling, Grover. 1988. *Strategies for Policy Making*. Chicago: Dorsey Press.

Stewart, Richard B. 1988. "Controlling Environmental Risks Through Economic Incentives." *Columbia Journal of Environmental Law* 13: 153–169.

Tietenberg, Thomas H. 1985. *Emissions Trading: An Exercise in Reforming Pollution Policy*. Washington, D.C.: Resources for the Future.

U.S. Environmental Protection Agency. 1990. *Reducing Risk: Setting Priorities and Strategies for Environmental Protection*. Washington, D.C.

Index

Designer: U.C. Press Staff
Compositor: Prestige Typography
Text: 10/13 Sabon
Display: Sabon
Printer: Haddon Craftsmen, Inc.
Binder: Haddon Craftsmen, Inc.